33 Advances in Polymer Science

Fortschritte der Hochpolymeren-Forschung

Edited by H.-J. CANTOW, Freiburg i. Br. · G. DALL'ASTA, Colleferro
K. DUŠEK, Prague · J. D. FERRY, Madison · H. FUJITA, Osaka
M. GORDON, Colchester · W. KERN, Mainz · S. OKAMURA, Kyoto
C. G. OVERBERGER, Ann Arbor · T. SAEGUSA, Kyoto · G. V. SCHULZ, Mainz
W. P. SLICHTER, Murray Hill · J. K. STILLE, Fort Collins

With 55 Figures

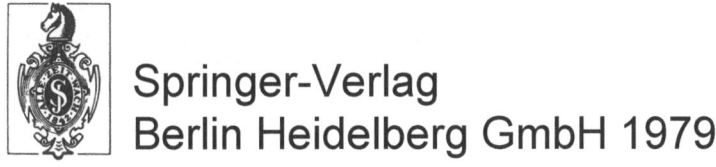

Springer-Verlag
Berlin Heidelberg GmbH 1979

Editors

Prof. Dr. Hans-Joachim Cantow, Institut für Makromolekulare Chemie der Universität, Stefan-Meier-Str. 31, 7800 Freiburg i. Br., BRD

Prof. Dr. Gino Dall'asta, SNIA VISCOSA – Centro Studi Chimico, Colleferro (Roma), Italia

Prof. Dr. Karel Dušek, Institute of Macromolecular Chemistry, Czechoslovak Academy of Sciences, 162 06 Prague 616, ČSSR

Prof. John D. Ferry, Department of Chemistry, The University of Wisconsin, Madison, Wisconsin 53706, U.S.A.

Prof. Hiroshi Fujita, Department of Polymer Science, Osaka University, Toyonaka, Osaka, Japan

Prof. Manfred Gordon, Department of Chemistry, University of Essex, Wivenhoe Park, Colchester C04 3 SQ, England

Prof. Dr. Werner Kern, Institut für Organische Chemie der Universität, 6500 Mainz, BRD

Prof. Seizo Okamura, No. 24, Minami-Goshomachi, Okazaki, Sakyo-Ku, Kyoto 606, Japan

Prof. Charles G. Overberger, Macromolecular Research Center, Institute of Science and Technology, The University of Michigan, Ann Arbor, Michigan 48 104, U.S.A.

Prof. Takeo Saegusa, Department of Synthetic Chemistry, Faculty of Engineering, Kyoto University, Kyoto, Japan

Prof. Dr. Günter Victor Schulz, Institut für Physikalische Chemie der Universität, 6500 Mainz, BRD

Dr. William P. Slichter, Chemical Physics Research Department, Bell Telephone Laboratories, Murray Hill, New Jersey 07 971, U.S.A.

Prof. John K. Stille, Department of Chemistry, Colorado State University, Fort Collins, Colorado 805 23, U.S.A.

ISBN 978-3-662-15416-8   ISBN 978-3-540-35229-7 (eBook)
DOI 10.1007/978-3-540-35229-7

Library of Congress Catalog Card Number 61-642

© by Springer-Verlag Berlin Heidelberg 1979

Originally published by Springer-Verlag Berlin Heidelberg New York in 1979
Softcover reprint of the hardcover 1st edition 1979

2152/3140 – 543210

* 26. 2. 1903   † 2. 5. 1979

Nobelprize in Chemistry 1963

Contents

Feasibility of Polymer Film Coatings Through Electroinitiated Polymerization in Aqueous Medium

Giuliano Mengoli

Laboratorio di Polarografia ed Elettrochimica Preparativa del C.N.R., Corso Stati Uniti 10, I-35100 Padova, Italy, in collaboration with Centro Ricerche Fiat, I–10043 Orbassano (Torino), Italy

The paper deals with the feasibility of polymeric coatings onto conductive substrates by electropolymerization as a new technique of metal protection. Although in theory the electropolymerization might provide by only one stage preparation of the polymer, spreading over the substrate and possibly crosslinking to insoluble networks, the situation is not so favorable. As a matter of fact, reviewing the work performed so far reveals that the polymerization "in situ" of traditional acrylic monomers, when possible, involves high current consumption and leads to unsatisfactory coatings. The reasons of such results are interpreted on the basis of both low production of radical initiator at the electrode and desorption phenomena: thus the most promising routes for further investigation are outlined.

Table of Contents

1 Introduction

The leading role for the protection against corrosion of metal sheets utilized for structures (i.e. car bodies, panels for electric household appliances) is now played by paint formulations based on synthetic polymers, which are generally applied to the substrate by electrophoretic deposition.

However, the electrophoretic coating technique[1-3] that has spread over the world since the 1960s due to its undoubted advantages demands a not insignificant energy supply: in fact high potential differences (100–150 V) are necessary for migration and deposition of the polymer particles and the resulting coatings must be subsequently cured at temperatures near 200 °C. A frequent fault of the organic films formed this way is the poor adhesion to the metal sheet: this is a consequence of the process features involving the flocculation of polymer micelles onto the substrate, without any previous interaction on a molecular scale between polymer and metal.

In the field of the electrolytic deposition of organic coatings, the electropolymerization "in situ" has lately become the object of some investigation[4]. In theory this method might provide by only one stage the preparation of the polymer, its spreading over the substrate and possibly its crosslinking to form insoluble networks.

In comparison with the electrophoretic coating, high electric potentials are not necessary: furthermore, as the electropolymerization is likely to take place "in situ" only when there is some specific interaction between monomer and metal electrode[5], a better adhesion for the resulting films should be generally expected.

As will be shown below, the aprotic medium offers an easier approach to the electropolymerization "in situ", but the possible technological interest of the technique may be assessed only when substituting water based compositions for expensive and polluting solvents like N,N-dimethylformamide (DMF), dimethyl sulfoxide (DMSO) or chlorinated hydrocarbons.

The feasibility of polymer film coatings by electro-initiated polymerization in water medium is discussed in this paper with reference to the following points:
The apparatus and procedure suitable for this type of investigation are briefly described.
Some restrictions imposed by water as solvent are provisionally identified.
The literature works on the subject up to this date are reviewed with the support of data obtained in the author's own laboratory.
Finally, the limits inherent in the technique are discussed, and the most promising avenues for further investigation are outlined.

2 Experimental Approach

Figure 1 illustrates an electrolytic cell which may be utilized satisfactorily for obtaining polymeric coatings onto metal sheets by electropolymerization in water.

It consists of a simple glass cylinder wherein the metal sheet (a) is the working electrode and the Pt coil (b) is the counter electrode. The cell cover (c) bears a slide-

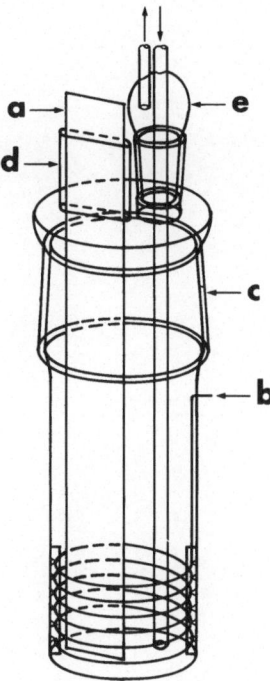

Fig. 1. Electrolytic cell for coating metal sheets by electropolymerization processes

way (d) to introduce the metal sheet or to withdraw it once coated: several sheets
may be consequently coated very easily one after the other in the same solution.
The inlet outlet equipment (e) allows, when required, to carry out the process in
nitrogen atmosphere.

The electrolyses are performed by applying a constant potential difference
(from 3 to 10 V) between the electrodes, the current variations during each run being monitored by an ammeter. The metal sheet is made either the cathode or the
anode, depending on what electrode reaction promotes the polymerization in situ
of the system under investigation: the counter electrode reaction is in any case supported by water discharge, the products of which do not significantly affect the in
situ reactions.

Electrolyzing by constant (small) potential differences applied generally leads
to obtain both a regular smooth formation of the polymer and a good throwing
power for the process when this is accomplished (as in the favorable instances) by
insulation of the metal sheet.

In view of these criteria, other electrolysis techniques seem less suitable for electropolymerization in situ: strong potential increases breaking the insulating films
could in fact arise from amperostatic electrolysis, whereas potentiostatic conditions
(if attainable at large electrodes during the formation of a coating) involve experimental complications without giving practical advantages.

The potentiostatic electrolysis is, however, a useful preliminary research tool:
Fig. 2 illustrates a compartmented cell suitable for testing on microscale, by steady
state potentiostatic technique, the electrochemical features of a given system and
the feasibility of macroscale coatings. (W), a metal rod of Fe or Zn, Cu, or Al sealed

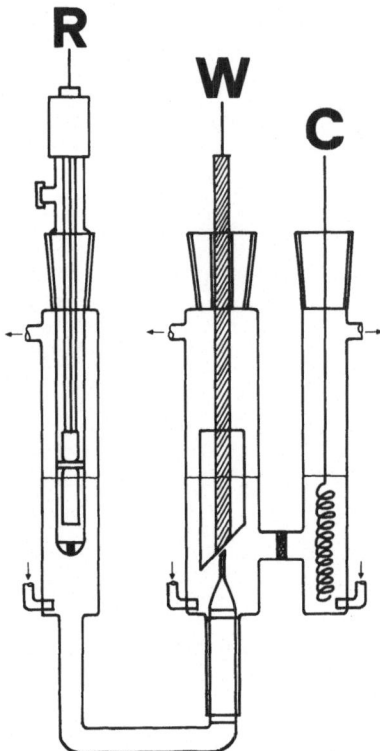

Fig. 2. Electrolytic cell for voltammetric runs

into a "Teflon" sheat having an exposed area 0.2 cm^2, is the working electrode; (C) is a Pt counter electrode; (R) is a standard calomel electrode (S.C.E.) separated from the working electrode by a Luggin capillary and a tap.

3 Limits of the Water Medium

The number of monomers exploitable for electropolymerization in water is first of all restricted by solubility limits. Two rather obvious ways of overcoming this handi-cap are:

Selecting a suitable buffer wherein monomers bearing acid or basic groups are dissolved more easily.

Adding to the medium some per cents of an alcohol. Methanol is particularly suitable to this end, having good affinity for several monomers and incompatibility with the corresponding polymers: at the same time the CH_3OH molecule is unlikely to affect significantly the adsorption of the monomers on the metal electrode.

The dissolution or dispersion of a monomer by a surface active compound does not seem an advisable route since the storage of the surfactant also in the double layer region might inhibit the polymerization in situ: on the other hand if a polymer coat-ing is formed, its features are likely to be degraded by the entrapped surfactant.

Table 1. Electrode processes leading to the formation of free radical intermediates as possible polymerization initiators (M = monomer)

Cathodic processes	
a) $H^+ + e \longrightarrow H^{\cdot}$ $\quad H^{\cdot} + M \longrightarrow M^{\cdot}$	proton discharge
b) $M + e + H^+ \longrightarrow M^{\cdot}$	direct monomer reduction
c) $Fe^{3+} + e \longrightarrow Fe^{2+}$ $\quad Fe^{2+} + H_2O_2 \longrightarrow Fe^{3+} + {\cdot}OH + {}^-OH$ $\quad {\cdot}OH + M \longrightarrow M^{\cdot}$	activation of a redox catalyst
d) $R_2O_2 + e \longrightarrow RO^{\cdot} + RO^-$ $\quad RO^{\cdot} + M \longrightarrow M^{\cdot}$	direct reduction of a peroxy compound
Anodic processes	
e) $Fe \longrightarrow Fe^{2+} + 2e$ $\quad Fe^{2+} + H_2O_2 \longrightarrow Fe^{3+} + {\cdot}OH + {}^-OH$ $\quad {\cdot}OH + M \longrightarrow M^{\cdot}$	anodic dissolution in presence of peroxy compounds
f) $CH_3COO^- - e \longrightarrow {\cdot}CH_3 + CO_2$ $\quad {\cdot}CH_3 + M \longrightarrow M^{\cdot}$	Kolbe reaction

A second limit imposed by working in water regards the mechanism of the polymerization: differently from the electropolymerizations in situ performed in aprotic medium which may propagate by ionic mechanisms[5-10], only free radical or stepwise polymerization are likely to be promoted in water. Consequently one has to select the monomers and to find out the electrode reactions suitable for a radical polymerization.

To this end Table 1 summarizes the electrode processes which are commonly estimated to initiate free radical polymerizations. These literature data are related to systems giving polymeric products in solution, utilizing inert Hg*, Pt or carbon electrodes: thus a first step is to determine the feasibility of such processes at metal electrodes as Fe or Zn, Cu, Al the coating of which is most important technologically.

Figure 3 shows the polarization curves obtained from an acid buffer (pH = 2) and Fig. 4 shows those resulting in alkaline medium (pH = 9.8).

The cathodic limit is given by the reduction of either protons or H_2O and the available range of potentials follows the hydrogen over-potential scale of the metals: the anodic limit is given by the oxidation of either H_2O or the metal electrode. When considering Figs. 3 and 4 with respect to the reactions of Table 1, it is clear that the use of reactive metal electrodes such as Fe, Zn, Cu, Al restricts the number of the electrode reactions which are within the reach of metals of higher inertness such as Hg for the cathodic and Pt for the anodic processes.

In fact: *Reaction a)*, initiation through H$^{\cdot}$ atoms may occur here at every metal. With respect to the formation of a protective coating reaction a) however, being always paralleled by H_2 evolution, may lead to gaseous occlusions in both the polymer

---

* Except for reaction e) of Table 1.

G. Mengoli

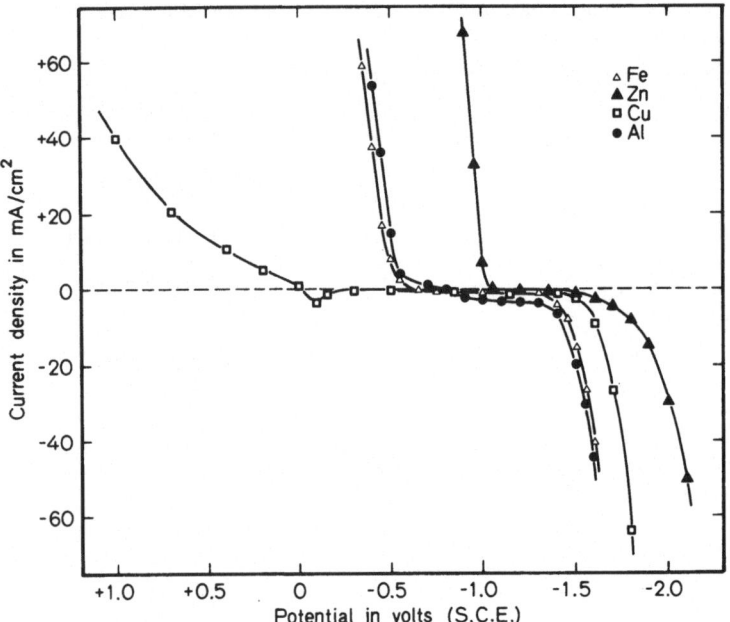

Fig. 3. Polarization curves recorded by steady state potentiostatic technique at electrodes of different metals as indicated in HCl/KCl buffer (pH = 2); S.C.E. = standard calomel electrode

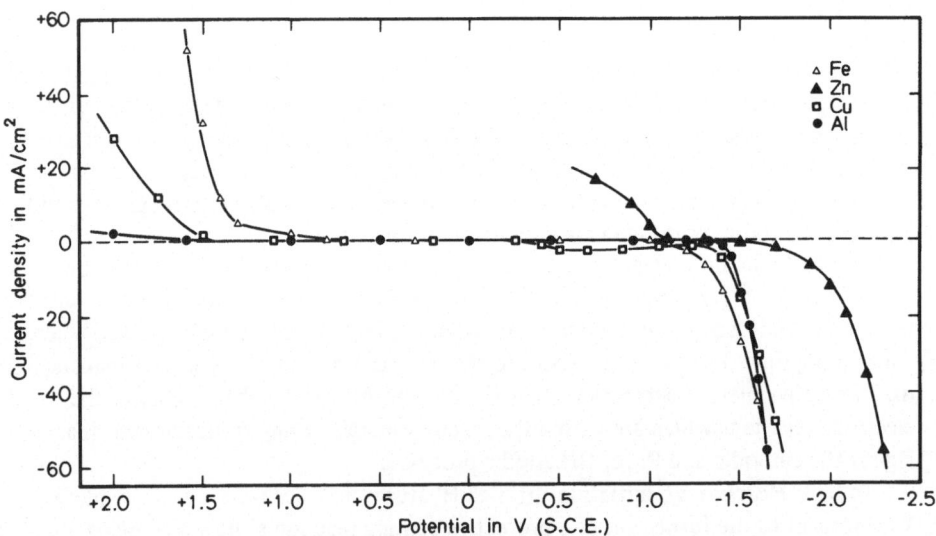

Fig. 4. Polarization curves recorded in KOH/KHCO₃ buffer (pH = 9.8) for different metal electrodes as indicated

film and the metal with consequent blistering of the first and brittleness of the second[11].

Reaction b), the direct monomer reduction may be realized at zinc, which has the highest hydrogen overpotential, but it is unlikely to take place at Fe electrodes before H$^+$ reduction.

Reactions c) and e) if fruitful, may be realized only at a particular metal electrode.

Reaction d), the reduction of peroxy compounds, especially when tried at transition metal cathodes, does not seem exploitable to initiate a polymerization as the radical intermediates are strongly adsorbed and quickly reduced before any possible interaction with a monomer[12].

Reaction f), the Kolbe reaction, is allowed here at copper probably, but CO_2 evolution might constitute a serious handicap to the polymer film formation.

4 Electropolymerization in situ of Acrylic and Vinylic Monomers

4.1 Polymerization of Diacetone Acrylamide

Diacetone acrylamide* was the first acrylic monomer to show the feasibility of polymer film coatings by electropolymerization in water medium. As a matter of fact the electropolymerization in situ of this monomer is favored by the following factors:

a) Diacetone acrylamide is soluble in water in a wide concentration range, whereas the corresponding polymer is insoluble: thus the polymerization once started at the surface of a conductive article is likely to proceed with the precipitation of the growing polymer chains onto the article.

b) Both carbonyl groups may be protonated with strong acids[13], thus allowing an easier diffusion and storing up of the monomer at the cathode.

c) The carbonyl group conjugated to the double bond may be directly reduced to free radical intermediates with consequent accumulation of initiating moieties at the cathode surface.

   The features of diacetone acrylamide polymerization are presented below.

4.1.1 Literature Data

Some patents assigned to the Grace company[14] are the first literature references claiming the cathodic polymerization of diacetone acrylamide as a method of protecting conductive articles. The examples reported therein illustrate very simple systems, generally comprising a 1 molar solution of diacetone acrylamide in diluted mineral acids (pH $\approx$ 1) which can be polymerized onto the surface of the article functioning as a cathode either by a direct current supply (4 mA/cm^2 for a potential difference applied of 20 V in a two compartment cell) or an alternating current; in this second instance both the electrodes can be coated during the cathodic half-cycle of the current.

---

* Systematic IUPAC nonmenclature: *N*-(1,1-dimethyl-3-oxobutyl)acrylamide.

With respect to the polymer film yield, the best coating performances were obtained on zinc, aluminium, lead, cadmium: however, the formation of thick films (≈ 1 mg/cm^2) always required more than 1 hour of electrolysis. The yields obtained on metals as Pt, Fe, Ni were very much poorer.

It was claimed that by increasing the potential difference between the electrodes the yield increases: the same result can be obtained by adding catalytic amounts of peroxy compounds.

Some of these results were confirmed by Jostan et al.[15-16]: they reported that diacetone acrylamide cannot be polymerized by either amperostatic or potentiostatic electrolysis, whereas the electropolymerization is successful when applying constant potential differences between the electrodes.

Working with aluminium cathodes, by a voltage = 12 V in a divided cell, it was found that the polymerization has three main periods:
a) an induction period in which no coating is formed;
b) a period in which the aluminium is rapidly coated;
c) a period of thickening of the coating at a lower rate.

The current sharply decreased from the first to the second period while it kept an almost stationary value during the third (the current went from 10 mA/cm^2 to a final value of 5 mA/cm^2).

Once corrected for the induction, the coulombic yield of the polymer film proved to be linear with the charge supplied, but the yield decreased and the coating was degraded when the electrolyses were performed with higher voltages.

No further light was eventually thrown on the kinetics and mechanism of diacetone acrylamide electropolymerization by a report by Oliver[17]: a technological interest was however claimed for the polymer films prepared this way, which when cured at 180–200 °C would become thermoresistant and completely insoluble in any solvent.

4.1.2 More Recent Results

Some further investigation on the electropolymerization in situ of diacetone acrylamide has been performed by the author of this review.

When the voltammetric behavior of an acid solution of diacetone acrylamide was examined by steady state potentiostatic technique, no formation of passivating polymer films onto a Fe microcathode could be tested since the current-potential curves obtained in the presence of monomer do not differ from those typical for the medium. Figure 5 conversely shows that at a Zn microcathode a remarkable inhibition for the reduction of the solvent takes place on addition of diacetone acrylamide. Following this preliminary experiment, the coating of Zn sheet cathodes was systematically studied to determine the influence of the reaction variables and to explain the features of the process.

a) Effect of the Potential Difference Applied. Once a potential difference sufficient to realize the polymerization was applied, the decrease of the current with the electrolysis time (i.e. the passivation rate of the Zn sheet cathode), was approximately assumed to be a test of the coating process efficiency.

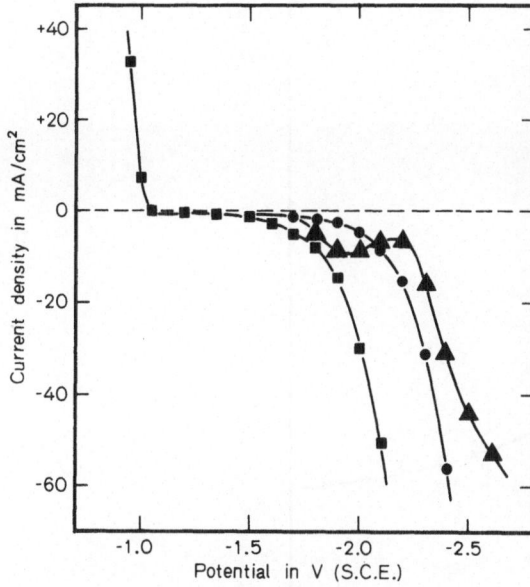

Fig. 5. Current-potential curves obtained at a Zn electrode in acid buffer (pH = 2) for various diacetone acrylamide concs.: ■ 0.00 M; ● 0.06 M; ▲ 0.30 M

Thus it was found that on electrolysing diacetone acrylamide 1 M in H_2SO_4 0.1 N, once exceeded a threshold potential difference ($<$ 3.5 V), the current drops with time as a consequence of formation and thickening of an insulating coating: the lowest value attained by the current corresponding to the most finished coating. According to Fig. 6, further increases of the voltage do not seem to increase the

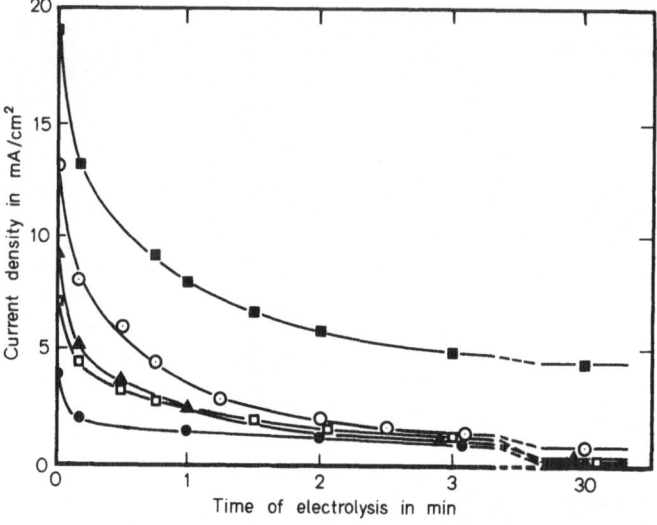

Fig. 6. Current decrease (i. e. passivation rate of a Zn sheet) during the electropolymerization in situ of diacetone acrylamide 1 M in H_2SO_4 0.1 N for increasing potential difference applied: ● 3.5 V; □ 4.0 V; ▲ 4.5 V; ⊙ 5.0 V; ■ 6.0 V

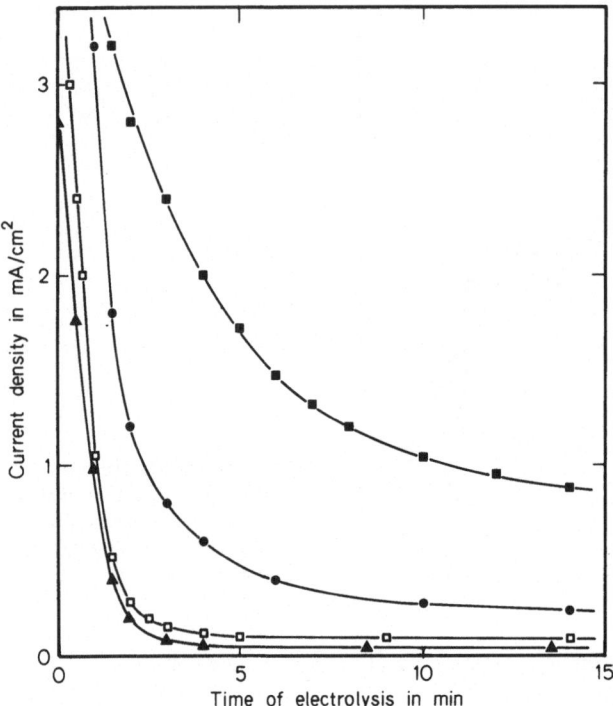

Fig. 7. Passivation rate of a Zn sheet at fixed diacetone acrylamide conc. and voltage applied, but for increasing H_2SO_4 concs.; H_2SO_4: ▲ 10^{-2} N; □ $2 \cdot 10^{-2}$ N; ● 10^{-1} N; ■ $2 \cdot 10^{-1}$ N

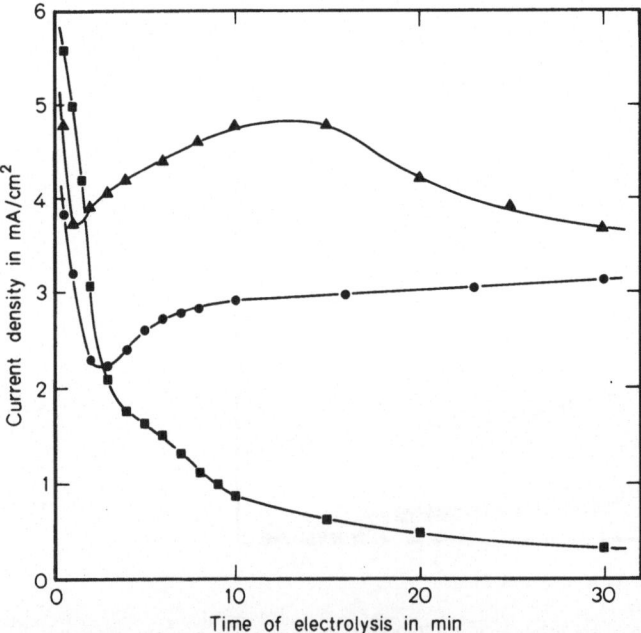

Fig. 8. Current trend during the electrolysis of diacetone acrylamide 1 M in H_2SO_4 0.2 N for various voltages: ▲ 3.75 V; ● 3.5 V; ■ 4.0 V

rate of polymer film formation: over 5 V the current attains high stationary values and with 7–8 V the coating cannot be built up anymore.

When the coatings obtained this way were examined, not only was the resulting morphology far better but also the coulombic yield (mg/coulombs) (see Table 2) was tremendously higher for the samples obtained with small potential differences applied: as an instance the yield, which was about 300 g/F(araday) at 3.5 V was reduced to ≈ 170 g/F at 5 V and to an order of magnitude below ≈ 10 g/F at 6 V.

b) Influence of pH. Figure 7 shows the Zn sheet cathode passivation rates, which result on electrolyzing diacetone acrylamide 1 M solutions containing different H_2SO_4 concentrations with 5 V. It can be observed that a decrease of the acid electrolyte concentration gives rise to a more efficient coating process (see also the data of Table 2). Not only the passivation rate depends on the pH, but also the threshold voltage necessary to start the polymerization. This fact clearly results when comparing the curves of Fig. 6 in H_2SO_4 0.1 N with those obtained in H_2SO_4 0.2 N of Fig. 8: the threshold voltage necessary for the polymerization reaches in the second case ≈ 4 V; no passivation of the Zn sheet cathode takes place for smaller potentials.

However, it must be noted that the presence of some low concentration of a strong acid favors the polymerization: in fact when a monomer solution of a neutral salt electrolyte was electrolyzed, the results were less satisfactory as may be deduced from the current trends illustrated by Fig. 9.

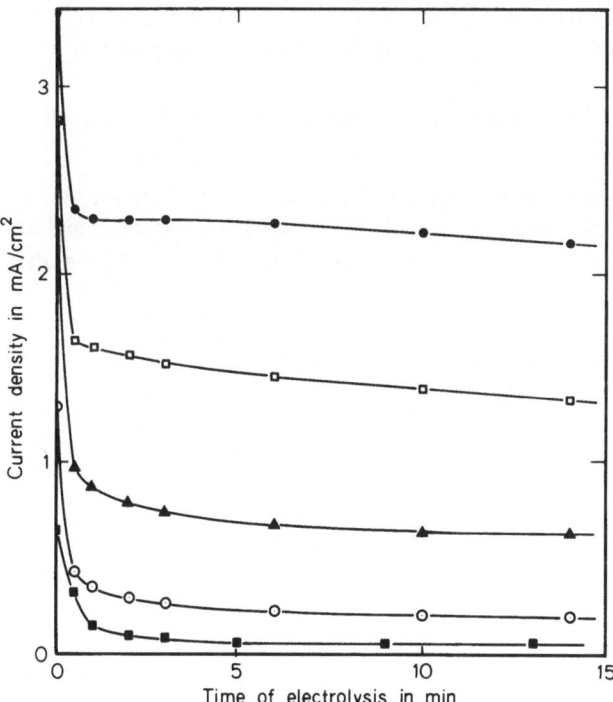

Fig. 9. Current trend during the electrolysis of diacetone acrylamide 1 M in Na_2SO_4 10^{-2}M (pH = 3.5) for increasing voltages applied: ■ 3.0 V; ⊙ 3.5 V; ▲ 4.0 V; □ 4.5 V; ● 5.0 V

Table 2. Coulombic yield of diacetone acrylamide in situ electropolymerization for various
experimental conditions[a]

Monomer conc.	H_2SO_4 conc.	Voltage in V	Yield in mg/cm^2	Yield in mg/Coulomb
1.0 M	0.01 N	5.0	0.74	3.12
1.0 M	0.01 N	6.0	0.80	1.79
1.0 M	0.02 N	5.0	1.23	2.80
1.0 M	0.10 N	4.5	1.56	1.85
1.0 M	0.10 N	5.0	1.80	1.68
1.0 M	0.10 N	6.0	1.12	0.11
0.5 M	0.10 N	5.0	0.65	0.23
1.5 M	0.10 N	5.0	2.41	2.50
1.0 M	0.20 N	5.0	1.80	1.00

[a] Temp. $t = 25\,°C$; Zn cathode; 30 min of electrolysis

   c) Effect of Monomer Concentration. On keeping at a fixed value voltage and pH,
both the passivation rate of Zn sheet cathode and the weight of polymer formed
thereon increase appreciably with the increase of monomer concentration. The second fact is shown by the kinetics of Fig. 10, where the yield of polymer in mg/cm^2
is plotted as function of the respective electrolysis times for three different diacetone acrylamide concentrations. Other than the polymerization rate, also the coulombic yield increases considerably with monomer concentration (see Table 2). In
Fig. 11 where the yield of polymer in mg/coulombs resulting from the previous
kinetics is plotted one can see that the coulombic yield is not constant in the deposition process: it grows from the beginning till it reaches a steady value, during a time
which may be relatively large.

   d) Molecular Weights. The molecular weights of the coatings obtained for the
various reaction conditions are reported in Tab. 3: the products are oligomers and do
not differ from each other substantially on varying the electropolymerization conditions.

Table 3. Molecular weight $\bar{M}_n$ of poly(diacetone acrylamide) samples
obtained by electropolymerization onto Zn sheets for various experimental conditions

Monomer conc.	H_2SO_4 conc.	Voltage in V	$\bar{M}_n$[a]
1.5 M	0.10 N	5	1150
1.0 M	0.10 N	5	1170
0.5 M	0.10 N	5	1360
1.0 M	0.02 N	5	1690
1.0 M	0.02 N	4	990

[a] Number-average mol. wts. determined by osmometry in
     acetone at $40\,°C$.

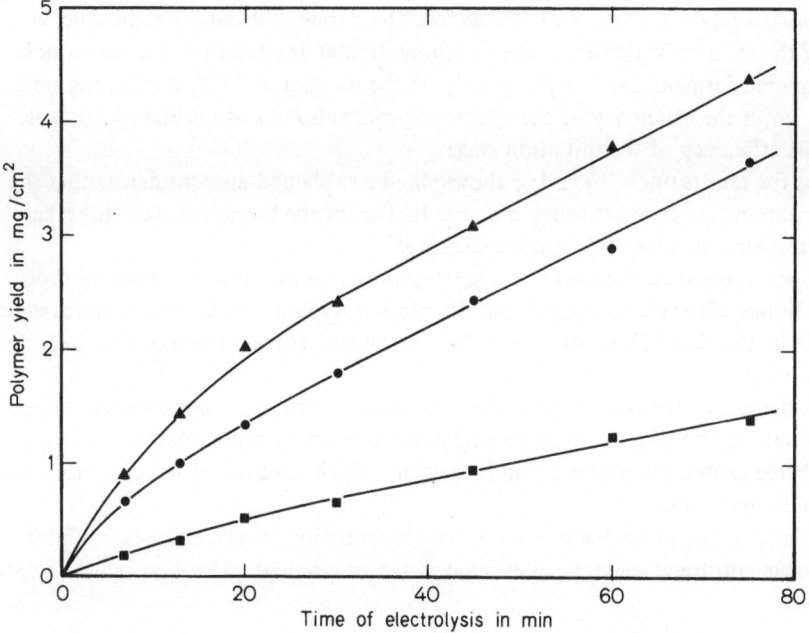

Fig. 10. Kinetics of diacetone acrylamide in situ electropolymerization in H_2SO_4 0.1 N with 5 V applied for various monomer concs.: ■ 0.5 M; ● 1.0 M; ▲ 1.5 M

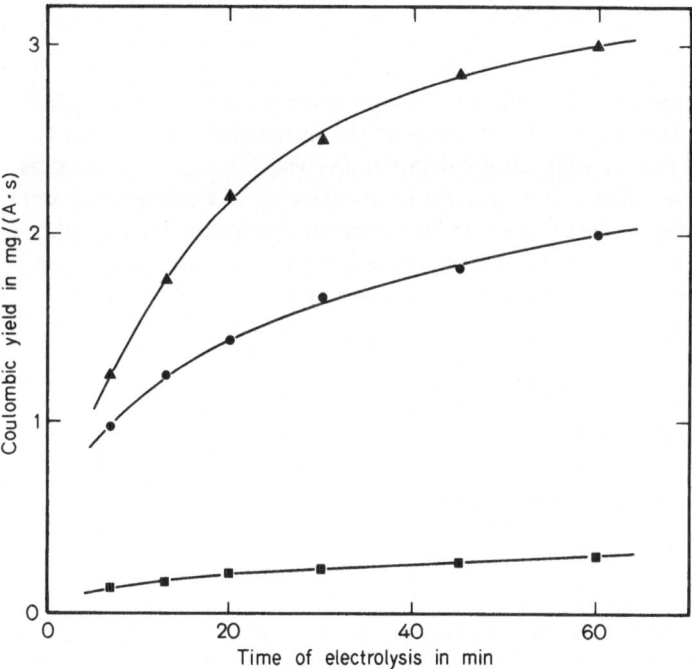

Fig. 11. Coulombic yields of the kinetics of Fig. 10 as function of the electrolysis time (Symbols as in Fig. 10)

e) Discussion on Kinetics and Mechanism. The strong molecular weight independence of the reaction variables allows to assume that at any time the amount of polymeric material formed in situ depends only on the number of fruitful initiating acts:
in other words the efficiency of the electropolymerization in situ is mainly controlled by the efficiency of the initiation stage.

Thus the results under b) and c) above may be explained on considering that the
polymerization is initiated through direct reduction of the monomer, but this reaction is paralleled by a parasitic proton discharge.

The polymerization threshold voltage dependent on pH, and the efficieny drop
at high H^+/monomer ratios suggest that diacetone acrylamide reduction requires some
over-potential so that below some pH value proton reduction is the more favored
cathodic process.

The increase of the coulombic yield with electrolysis time and monomer concentration resulting from Fig. 11 may be explained likewise by some inhibition to proton discharge caused by polymer film formation, which does not affect the efficiency
of monomer reduction.

On the other hand, as diacetone acrylamide reduction to a free radical initiator
requires some protonation step, an efficiency decrease at high pH values is also under-
standable.

$$M + H^+ \rightarrow MH^+ + e$$
$$\searrow$$
$$MH^.$$
$$\nearrow$$
$$M + e \rightarrow M^{\overline{.}} + H^+$$

Finally, with respect to the influence of the potential difference applied (cf. *a*)),
one has to consider that this variable affects both the current and the potential of
the Zn cathode (see Fig. 5). High current densities involving large material transfers
and convection are very likely to impede the regular growing of a polymer film onto
the substrate. Furthermore the more negative potentials assumed by the Zn cathode,
for increasing voltages applied, impede the polymerization in situ since they interfere
with monomer adsorption: this is a quite general phenomenon which will be discussed in more detail below.

4.2 Polymerization of Methyl Vinyl Ketone

Methyl vinyl ketone seems suitable for cathodic polymerization in situ from water
medium since it has good solubility in water whereas the polymer is insoluble. Furthermore the carbonyl group conjugated to the double bond might be reduced at
several metal electrodes with formation of radical initiators.

The situation for this monomer consequently seems to be very near to that for
diacetone acrylamide. However, no paper has appeared so far in the chemical literature on the electropolymerization in situ of methyl vinyl ketone. As a matter of fact
the indications which may be preliminarily obtained on the feasibility of such a pro-

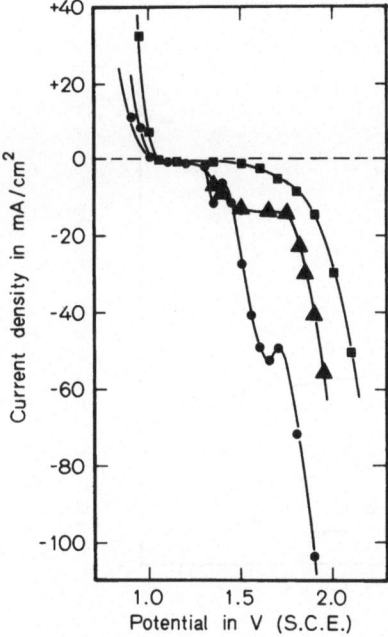

Fig. 12. Current-potential curves obtained at a Zn electrode in acid buffer (pH = 2) for various methyl vinyl ketone concentrations: ■ 0.00 M; ▲ 0.12 M; ● 0.6 M

cess are unfavorable. For instance, when the voltammetric behavior of methyl vinyl ketone at a Zn microcathode was investigated, no passivation but well defined reduction waves proportional to monomer concentration could be observed, as is illustrated by Fig. 12.

At a Fe microcathode on the other hand, some passivation became evident under the same conditions, but when the reduction was performed on a larger scale (in the cell of Fig. 1) the polymerization propagated also into the solution.

However, a more thorough investigation of the subject was subsequently able to show that methyl vinyl ketone polymerization propagates also into the bulk of the solution with HCl, H_2SO_4, HNO_3 or their salts as electrolytes, whereas it develops in situ, confined to the Fe sheet cathode surface, when HF or KF are utilized. The results thus obtained may be summarized as follows: Figure 13 illustrates the variations of the current with the electrolysis time resulting when increasing potential differences are applied to aqueous solutions containing methyl vinyl ketone 1 M, F^- 0.2 M and having pH = 1. One can observe that the per cent decrease of the current is faster for the smallest potentials and indeed coherent and homogeneous coatings were obtained only under these electrolysis conditions (from 3 to 4 V). Therefore the influence of the potential applied on the coating process parallels that seen above for the diacetone acrylamide system.

Figure 14 shows the polymer film increase with electrolysis time for three different monomer concentrations: the linear relationship between yield and time (which is not maintained for the longest times) despite the current drop, depends probably on a lower coulombic yield of the polymerization during the initial time.

The coulombic yields (mg/coulomb) also for the best conditions of potential and pH, were always very poor since they were ranging from 0.05 to 0.4 with methyl vinyl ketone 0.9 M and 2.7 M respectively.

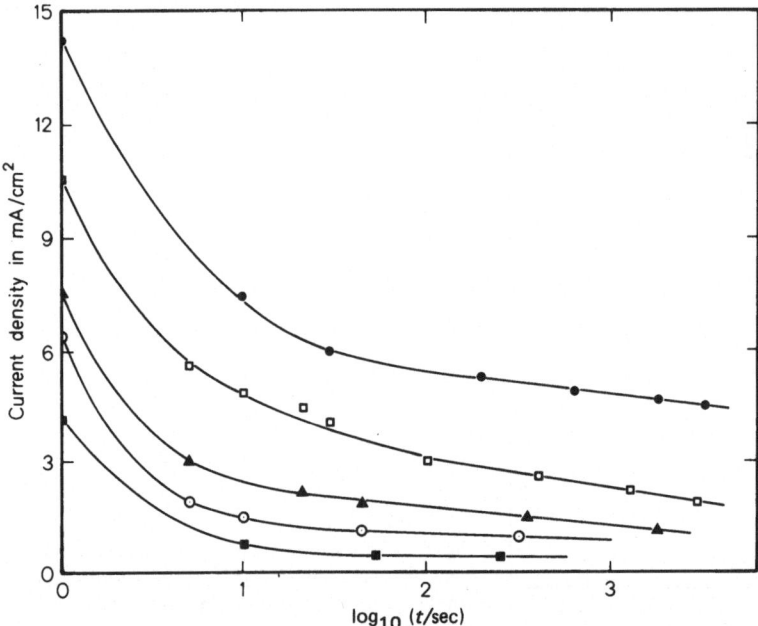

Fig. 13. Current decrease (i. e. passivation rate of a Fe sheet) during the electropolymerization time t of methyl vinyl ketone in situ for increasing voltages: ■ 3.0 V; ⊙ 3.3 V; ▲ 3.5 V; ◻ 4.0 V; ● 4.5 V

With respect to the molecular weight of the polymers, the $\overline{M}_n$ were about 2500, independent of the conditions of polymerization and the time scale of the process (see Table 4).

In conclusion, it can be said that the electropolymerization in situ of this monomer shows some similarities with the diacetone acrylamide system: a main difference lies in the initiation stage which takes place via proton discharge; at zinc in fact the

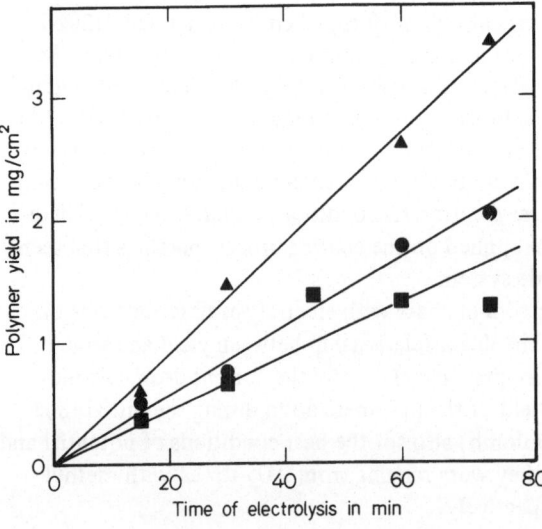

Fig. 14. Kinetics of methyl vinyl ketone in situ electropolymerization for various monomer concs.: ■ 0.9 M; ● 1.8 M; ▲ 2.7 M. (Voltage = 3.5 V)

Table 4. Molecular weight $\bar{M}_n$ of poly(methyl vinyl ketone) samples obtained by electropolymerization onto Fe sheets for various electrolysis times

Electrolysis time in min	$\bar{M}_n{}^a$
15	2720
30	2260
60	2310
90	2500

a   Number-average mol. wts. determined by osmometry in acetone at 40 °C. The polymers were formed in situ from a 2.7 M solution of the monomer containing $F^- = 0.2$ M and having pH = 1.

possible free radical intermediates of direct methyl vinyl ketone reduction are further reduced or dimerize before interacting with other monomers. Also the specific effect of F^- ions on keeping the polymerization in situ must be mentioned: with some speculation one might note that F^- ions probably have a positive free energy of adsorption[11] and thus they do not inhibit the storage of this monomer at the interface as is the case for anions of greater size.

4.3 Polymerization of Acrylamide

A peculiar type of electropolymerization in situ has been patented by Sobieski and Zerner[18]. These authors found that aqueous solutions of acrylamide (the main monomer), N,N'-methylene- or N,N'-ethylenebisacrylamide (the crosslinking agent), and $ZnCl_2$ (the catalyst) when electrolyzed in an undivided cell by a constant potential difference, in a few seconds yield a polyacrylamide film coating on the cathode surface, the bulk of the system being unaffected by the polymerization.

The process is applicable to several metal substrates, it reaches its optimum yield for a pH in the range 3.5–5 and the adhesion of the polymer to the substrate is favored by the presence in solution of catalytic amounts of Cu^{2+} ions or H_2O_2.

To explain the initiation mechanism of the process, it has been suggested that oxygen present in solution interacts with Zn^{2+} ions to form a reducible complex producing free radical initiators according to the overall reaction scheme:

$$Zn^{2+} + O_2 \longrightarrow ZnO_2^{2+}$$
$$ZnO_2^{2+} + e \longrightarrow ZnO_2^{+}$$
$$ZnO_2^{+} \longrightarrow Zn^{2+} + O_2^{-}$$
$$O_2^{-} + H_2O \longrightarrow HO_2^{\cdot} + {}^-OH$$
$$H_2O + HO_2^{\cdot} \longrightarrow H_2O_2 + HO^{\cdot}$$

The initiation mechanism of this polymerization has been reinvestigated more recently by Collins and Thomas[19], who found that the process takes place also when the solutions have been carefully freed from oxygen: thus an initiation through the cationic complex ZnO_2^{2+} seems unlikely.

On the contrary, evidences were presented which supports the formation of a cationic complex between acrylamide and zinc monohydroxide having the structure:

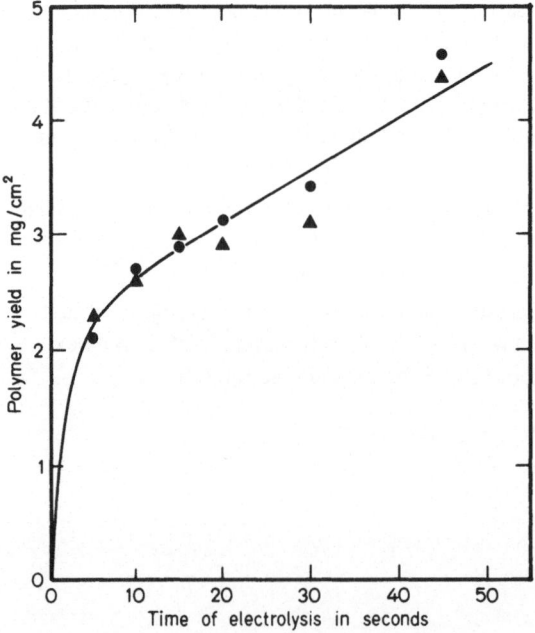

Since this complex is formed by donation of acrylamide carbonyl electrons to the Zn^+ cation, the carbonyl should become suitable of electrode reduction to a free radical intermediate: the interaction of this intermediate with the monomer stored at the cathode starts the polymerization process.

The data obtained from this system by the author of the present review may be summarized as follows: once applied a low potential difference, from 3 V on, between a Fe sheet cathode and a Pt anode (see the cell of Fig. 1) about 1 mg/cm² of homogeneous polymer film coating was formed in a few seconds on the Fe sheet, as results from Fig. 15. For longer electrolysis times the weight increase was rather slight and seemed to be independent of the potential difference applied since the same yields were obtained with different voltages. This fact is more clearly evidenced by Fig. 16 where the flat curve, almost parallel to the potential scale on the abscissa,

Fig. 15. Kinetics of acrylamide (1.4 M)/N,N'-methylenebis-acrylamide ($6.5 \cdot 10^{-2}$ M)/ $ZnCl_2$ (0.5 M) in situ electropolymerization. Voltages applied: ▲ 3.6 V; ● 4.6 V

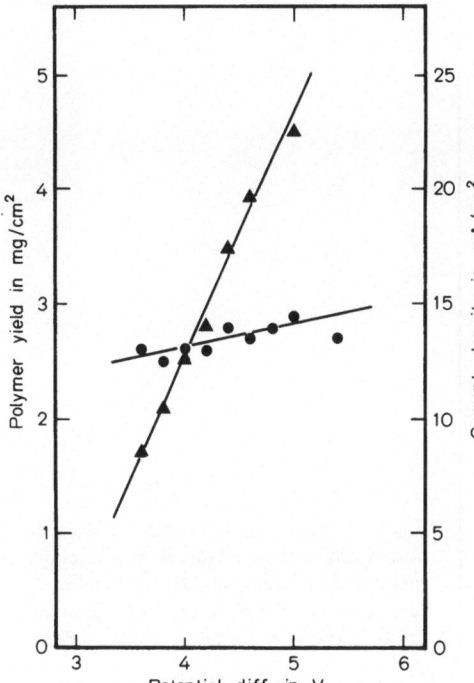

Fig. 16. Left-hand ordinate: polymer yield obtained from the system of Fig. 15 after 20 s for various voltages applied (•). Right-hand ordinate: variation of the current with the voltage (▲)

shows the polymer yields (left side ordinate) obtained after a fixed electrolysis time (20 s) for different potentials applied: the same flat curve shows that the polymer yields were independent also from the amount of current furnished to the system which increased with increasing voltage. As a matter of fact during coating formation the current did not drop from its initial value, and (with reference to the right side ordinate) Fig. 16 illustrates the linear relationship existing between potential and current at any time of the process.

These outstanding features may be explained as follows: acrylamide has a high affinity for the metal substrate and stores up at the metal-solution interface, consequently a polymerization which propagates mainly along the cathode surface (that is where the monomer concentration is higher) may be initiated by a catalyst formed in situ. Following the initiation, soon a stage is attained wherein the diffusion of the monomer to the metal sheet becomes the rate determining step of the process: this explains the independence of the yields from the voltage and the charge transferred.

The absence of any increase of ohmic resistance during formation of the coating is due to the high affinity of polyacrylamide chains for the protic medium: being highly crosslinked (by N,N'-methylenebisacrylamide) they do not dissolve but completely soak in the electrolytic solution. Although the coatings do not cause any shield to the electric field, the throwing power of the process is however good, since this polymerization in situ is mainly controlled by adsorption and diffusion phenomena which are not influenced by the low voltages utilized.

It must finally be mentioned that the formation of a complex between $ZnCl_2$ and acrylamide is here supported by the current-potential curves shown by Fig. 17.

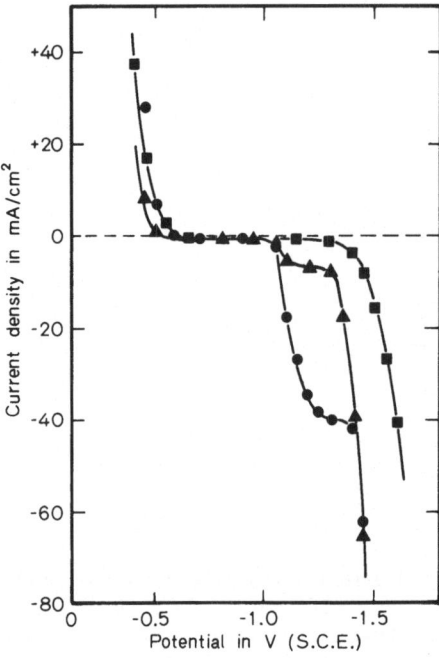

Fig. 17. Influence of acrylamide on the reduction of $ZnCl_2$ at a Fe cathode. ■: reduction curve of the buffer (pH = 2); ●: reduction curve of $ZnCl_2$ 0.5 N; ▲: reduction curve of $ZnCl_2$ in the presence of acrylamide 0.8 M

The reduction curve of $ZnCl_2$ is almost completely suppressed by the addition of acrylamide: since not passivating polymer film is formed on the cathode, the shift of the curve towards more negative potentials may be considered typical of a process whereby an easily reducible ion is complexed to form a less reducible species.

4.4 Polymerization of Other Monomers

The use of electropolymerization to coat metallic electrodes was investigated in detail by Subramanian et al.[20, 21], using a large number of monomers in either aprotic or aqueous media.

Thus it was found that for long electrolysis times (several hours) and considerable transfers of electric charge most of the monomers give polymeric deposits onto metals as Fe, Cu, Ni, Al etc.

Based upon their physical appearance these deposits were subdivided into the following three classes:

The first class comprises coatings having powdery to spongy appearance when removed from the cell; these poor features were kept by the dried coatings since a slight mechanical shock tended to separate powdery material from the substrate.

The second class consists of coatings having a planar surface appearance not much different from a painted coating: however they are swollen by the solvent and crack extensively upon drying. Such coatings were obtained by cathodic polymerization (maybe "via" an anionic mechanism) in DMF using generally acrylonitrile as monomer: through continuous electrolysis for several hours they could be made to grow up to 100 μm.

The third class of coatings, which are more pertinent to the topic of this review, were obtained in water containing H_2SO_4 as electrolyte by utilizing acrylic monomers such as acrylonitrile, acrylic acid or their mixtures. The polymerization was initiated via proton reduction on the cathode, by applying a constant potential difference of 8–10 V with the current density ranging from 10 to 15 mA/cm^2. The coatings consisted of very thin polymer films, as 1 μm of thickness (at the best) could be obtained after more than one hour of electrolysis: during their formation no remarkable current drop from the initial value was generally observed since the wet polymer film caused no electric resistance increase to the system: however the dried films have been found to give some protection to the metal substrate against chemical and electrochemical corrosion.

5 Limits of the Electropolymerization in situ

The data of the works by Subramanian et al.[20, 21] underlined above, emphasize further some limits which were intrinsic also for the more favorable instances relating to diacetone acrylamide or methyl vinyl ketone.

As a matter of fact, with the exception of the acrylamide/N,N'-methylenebisacrylamide/$ZnCl_2$ polymerization, which may be considered a "sui generis" system, it has been seen that:

a) considerable current consumption is required by the in situ electropolymerization process,

b) the electrolysis times are too long: any attempt to accelerate the polymerization rate by increasing the electric parameters (e.g. the potential difference applied) lead to both a drop in the polymer yield and a deterioration of the coating morphology,

c) the molecular weights of the polymer films are generally very low and require additional curing,

d) the throwing power of the process, especially at the highest potential differences applied, is not satisfactory as the polymer forming does not set a significant ohmic resistance against the electric field.

The limits depend mainly on the following factors:
the first one is the low efficiency of the electrodic generation of free radical initiators.

As a matter of fact it has to be noted that one of the most favorable examples seen above, that is diacetone acrylamide polymerization onto Zn sheet cathodes, required the passage of 1 Faraday for obtaining ≈ 300 g of polymer coating: on the basis of the $\overline{M}_n$ reported in Table 3, this yield corresponds to a current efficiency of $\approx 20\%$ only.

When the initiation takes place through proton discharge the situation is still more unfavorable: so on comparing the best current yields with the weight of the products for methyl vinyl ketone, it is found that a polymer molecule is produced by one fruitful electron transfer in more than fifty.

The second fact to be taken into account is monomer adsorption on the metal electrode, which is likely to be a necessary condition for the polymerization in situ.

It has been suggested that water[20, 21] because of its higher surface tension as compared to organic compounds is a unique solvent in favoring monomer adsorption to the metals and the adhesion of the polymers which may be obtained thereby. This is true only in part as the case is not always so favorable: for the metals of the Fe group, which clearly are the most important to protect against corrosion, the energy freed in the adsorption of water is considerably higher than for metals of high hydrogen overpotential and thus on Fe group metals organic adsorption is less easy[22].

Furthermore, also when the adsorption is favored by the presence of groups in the organic molecule such as carbonyl, amine, π-bonds (as really happens for most of the monomers) which may interact with, for instance, the d-orbitals of a transition metal, the adsorption takes place anyway within a well defined potential range, not far (for uncharged monomers) from the zero charge potential of the metal. It is known[11] that for increasing overpotentials away from the zero charge point, the organic molecules desorb and are replaced by H_2O molecules: therefore the overpotentials necessary to initiate the polymerization by some faradaic process may be well away from the zero charge potential of the metal (which is generally near to its static potential in the system) and consequently some monomer desorption is likely to take place. This may be often the main reason for the failure of the in situ electropolymerization for high potential differences applied, which involve increased over-potentials for the electrodes: any possible increase in the production rate of initiating species which should lead to a higher polymerization rate is vanished by the drop of monomer concentration at the metal-solution interface.

At this point one might object that adsorption is likely to take place only for a few layers thickness on the metal, while polymer film coatings some μm thick were presented so far: however, as a first polymer layer is built, a new interface is formed and the monomer will generally store up on its own polymer surface.

These considerations on the intrinsic limits of electropolymerization in situ from water medium seem to indicate that the technique when directly applied to the traditional acrylic and vinyl monomers is unlikely to become advantageous: but, a way may be found overcoming these limits once they are identified.

The polymerization of acrylamide/N,N'-methylenebisacrylamide/$ZnCl_2$ may be considered as one of such ways: here in fact the formation of a zinc-acrylamide reducible complex on the one side gives an efficient polymerization initiator and on the other side allows the cationic complexed monomer to be stored at the negatively charged metal surface. Unfortunately the physical characteristics of the coatings thereby obtained do not parallel the ease of their formation.

Other routes to be attempted are exemplified in the following section.

6 Promising Development Routes of the in situ Technique

6.1 Non Electrolytic Polymerizations in situ

It is possible to get over the limits discussed so far when an efficient production of free radical initiators is obtained on the surface of the metallic article without interfering with the adsorption of suitable monomers on the same surface. These conditions are realized by a technique the background of which may be seen in the following literature data:

I) a claim of previously cited patents on diacetone acrylamide[14] whereby the addition of H_2O_2 favors rates and yields of the polymerization process.

II) A patent by Hodes et al.[23] claiming the electropolymerization onto metallic cathodes (also of Fe) of acrylamide/N,N'-methylenebisacrylamide systems by reduction of H_2O_2 or other peroxides in acid medium.

III) The absence of any clear literature evidence (in sharp contrast to such claims) that the cathodic reduction of H_2O_2 may be a reliable source of free radicals. On the contrary, since the pioneering work by Kolthoff et al.[24] the possibility of polymerizing acrylic monomers by the sole electrolysis of peroxy compounds has been repeatedly denied[12].

IV) Finally a more recent patent by Hodes et al.[25] claiming the electropolymerization without current flow as a technique to obtain polymer coatings in an imagewise fashion on Fe surfaces. The process was realized by making a stainless steel plate the cathode of a light responsive cell, applying a potential of about 60–100 V across said cell for several seconds and causing imagewise activation of the metal plate: a polymer coating was then deposited on the activated area by immersing the plate in a strongly acid solution containing H_2O_2 and suitable acrylamide/N,N'-methylenebisacrylamide mixtures. Practically the sole function of the electrolytic activation was to remove oxide layers covering the plate so that the bare metal could react with the acid/peroxide/monomer system.

Therefore, from such results it was easy to think that the coatings obtained onto Fe cathodes during the electrolysis of diacetone acrylamide or acrylamide solutions containing H_2O_2 claimed by literature as formed by electropolymerization had been more likely obtained by the catalysis of the Fe^{2+}/H_2O_2 redox couple originated in situ from Fe corrosion. In this view the potential applied to the Fe sheet cathodes controlled only the metal dissolution in the acid medium to a low extent which did not cause any polymerization in the bulk.

As a matter of fact the polymerization of either diacetone acrylamide[26] or diacetone acrylamide derivatives[27] onto Fe articles "via" the corrosion of the metal in the presence of peroxy compounds has been claimed as a method of formation in situ of protective coatings: the feasibility of a similar process by utilizing other vinyl monomers in aqueous organic solvent solution has been claimed, too[28]. It must be noted, however, that the stability of such systems during repeated coating operations is rather low since the polymerization process often propagates also into the bulk of the solutions. These considerations prompted Mengoli et al.[29] to a more thorough investigation of the polymerization of acrylamide and some of its derivatives onto Fe plates "via" corrosion reactions in the presence of peroxy compounds. The following facts favoring such a process could be evidenced:

a) the corrosion of a Fe sheet (or Fe article) to Fe^{2+} assures an efficient initiation in situ: the corrosion may be promoted by either H^+, or other suitable cations, or the same peroxy compounds: Table 5 resumes some of the possible reactions taking place at a Fe surface using H^+, Cu^{2+} and Fe^{3+} ions as corrosion agents. One has to think that such type of initiation might work also at the surface of

Table 5. Initiation mechanism by corrosion processes

1) Fe $\rightarrow Fe^{2+} + 2e$ anodic
2) $2H^+ \quad + 2e \quad\;\; \rightarrow H_2$ cathodic
3) $Fe^{2+} \quad + ROOH \;\; \rightarrow Fe^{3+} + RO^{\cdot} + {}^{-}OH$
4) $2Fe^{3+} \;\; + Fe \quad\;\;\; \rightarrow 3Fe^{2+}$
5) $Cu^{2+} \quad + Fe \quad\;\;\; \rightarrow Cu \quad + Fe^{2+}$
6) $Cu^{2+} \quad + Cu \quad\;\;\; \rightarrow 2Cu^+$
7) $Cu^+ \quad\;\; + ROOH \;\; \rightarrow Cu^{2+} + RO^{\cdot} + {}^{-}OH$
8) $Fe^{3+} \quad + Cu \quad\;\;\; \rightarrow Fe^{2+} + Cu^+$

    metals other than Fe, which may dissolve as oxidized forms able of interacting with peroxy compounds.

b) During the corrosion, the Fe sheet acquires a potential of ca. -0.5 V against standard calomel electrode (S.C.E.) probably near to the zero charge potential of the metal[22]: therefore monomer adsorption on the plate should be at a maximum.

c) The polymerization may be confined to the surface of the metallic sheet, without affecting the bulk of the system by keeping in solution a constant O_2 (or air) concentration which acts as inhibitor.

    Thus, by utilizing a cell similar to that of Fig. 1 (clearly, without applying any external potential difference) and bubbling into the monomer solution a slight O_2 draft, the kinetics of either diacetone acrylamide or acrylamide/N,N'-methylene-bisacrylamide in situ polymerizations could be investigated by dippling several sheets in succession into the same reaction system for different interaction times. It was

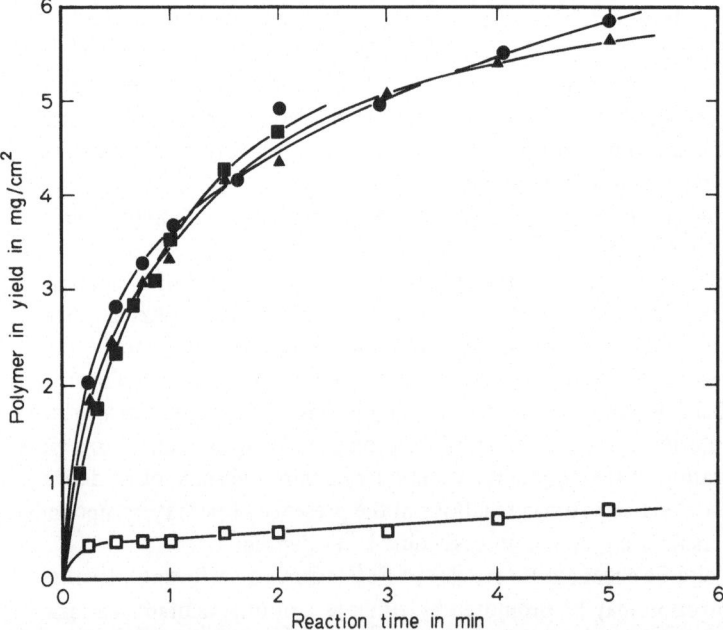

Fig. 18. Kinetics of diacetone acrylamide polymerization onto Fe sheets for different pH's. pH equal to: □ 2.5; ■ 2.0; ▲ 1.0; ● 1.5. Monomer conc. = 0.59 M; *tert*-butyl hydroperoxide = $0.88 \cdot 10^{-3}$ M; temp. $t = 25\,°C$

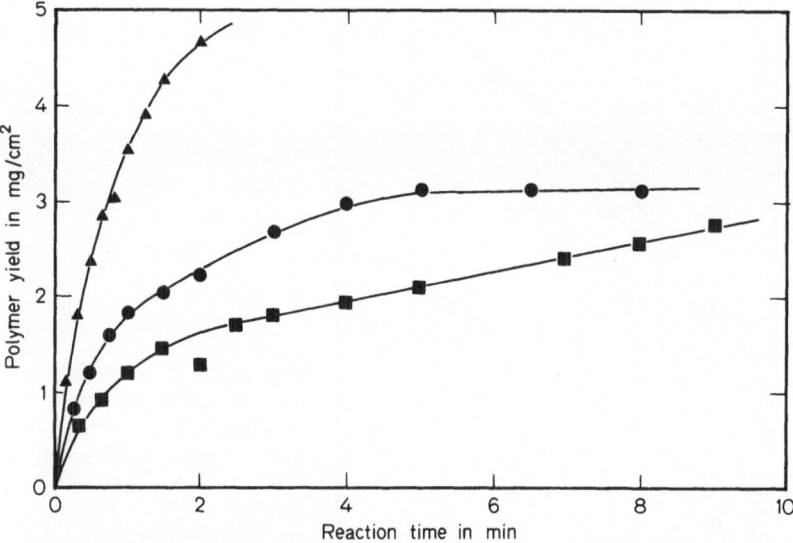

Fig. 19. Kinetics of diacetone acrylamide polymerization in situ for different monomer concs: ■ 0.17 M; ● 0.29 M; ▲ 0.59 M

found that the process may be realized only within a well defined pH range (from ≈1 to ≈4) and the rate of in situ polymer formation, that is the thickness increase of the coating can be controlled by acting on one of the following variables: reaction time, monomer concentration and, although at a lower extent, peroxide concentration. For instance Fig. 18 shows the kinetics of diacetone acrylamide polymerization obtained for different pH's; Fig. 19 illustrates the effect of monomer concentration and Fig. 20 eventually shows the kinetics of acrylamide/N,N'-methylenebisacrylamide polymerization initiated with different $K_2S_2O_8$ concentrations.

Table 6. Molecular weight $\bar{M}_n$ of poly(diacetone acrylamide) samples obtained on Fe plates for different experimental conditions[a]

Monomer conc.	$tert$-Butyl hydroperoxide conc.	pH (HCl)	$\bar{M}_n$
5.90 M	$6.6 \cdot 10^{-3}$ M	2.0	3650
2.95 M	$6.6 \cdot 10^{-3}$ M	2.0	3650
1.77 M	$6.6 \cdot 10^{-3}$ M	2.0	4500
5.90 M	$6.6 \cdot 10^{-3}$ M	2.5	4750
5.90 M	$6.6 \cdot 10^{-3}$ M	1.5	3600
5.90 M	$6.6 \cdot 10^{-3}$ M	1.0	2200
5.90 M	$13.2 \cdot 10^{-3}$ M	2.0	4200
5.90 M	$2.2 \cdot 10^{-3}$ M	2.0	2850
5.90 M[b]	$2.2 \cdot 10^{-3}$ M	2.0	2400
5.90 M	$0.88 \cdot 10^{-3}$ M	2.0	2700

[a] The number-average mol. wts. were determined in acetone at 40 °C by osmometry.
[b] The deposition was performed from a solution containing $CuCl_2 = 8.8 \cdot 10^{-3}$ M.

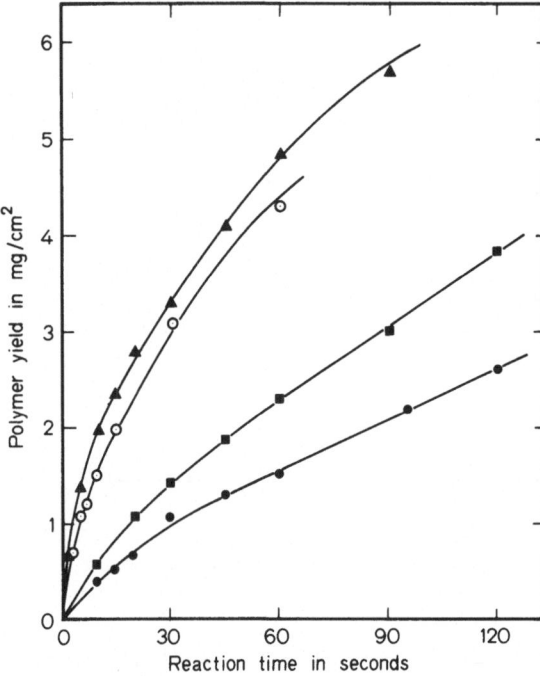

Fig. 20. Kinetics of acrylamide (1.4 M)/N,N'-methylenebis-acrylamide ($6.5 \cdot 10^{-2}$ M) polymerization in situ for different $K_2S_2O_8$ concs. with $CuCl_2$ ($8.8 \cdot 10^{-3}$ M) as corrosion agent of Fe; pH = 1.2; $K_2S_2O_8$ equal to:
● $0.5 \cdot 10^{-2}$ M;
■ $0.83 \cdot 10^{-2}$ M;
⊙ $1.66 \cdot 10^{-2}$ M;
▲ $3.3 \cdot 10^{-2}$ M

Regarding the physical characteristics, it has to be noted that poly(diacetone acrylamide) coatings obtained this way showed higher $\overline{M}_n$ than the corresponding products obtained by electrolysis on Zn sheets (see Tab. 6): this fact leads to think that some electrolytic termination as:

$$\mathsf{\Lambda\!\Lambda\!\Lambda} \overset{|}{\underset{|}{C}} \cdot + e + H^+ \rightarrow \mathsf{\Lambda\!\Lambda\!\Lambda} \overset{|}{\underset{|}{C}} H$$

must be taken into account in the second case.

This type of investigation was extended to other monomers, reaching favorable results for some acrylic acid monomers[30]: for the same conditions methacrylic acid derivatives did not polymerize in situ, thus likely emphasizing the importance of steric factors on the process.

6.2 Electropolymerization in situ of Phenols

Another promising research field is that concerning monomer systems having some oxidation or reduction step in their polymerization which is suitable to be performed electrochemically.

This is the case of phenols which, as is known[31], may undergo an oxidative coupling polymerization to polyoxyphenylenes.

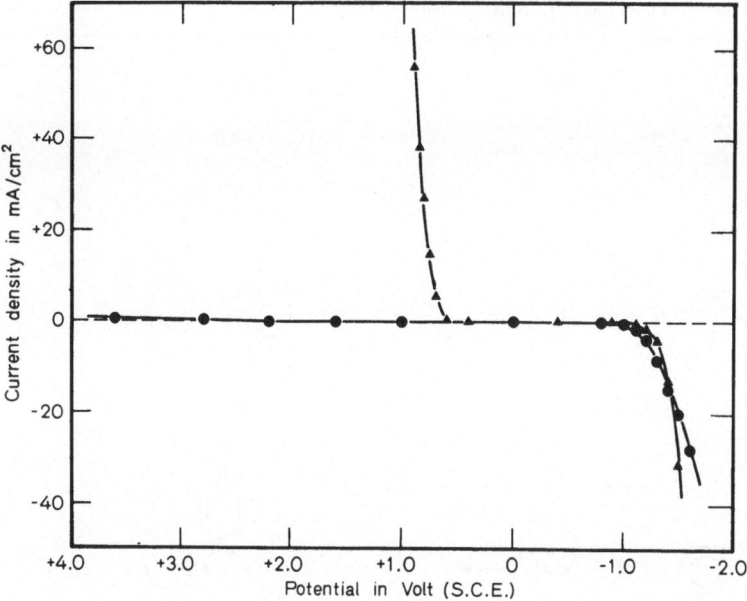

Fig. 21. Passivation of a Fe anode in NaOH solution caused by phenol: ▲ in the absence of phenol; ● in the presence of phenol 0.5 M

With reference to electrochemical polymerization, Fig. 21 shows the current-potential curve for the oxidation of a Fe microanode in alkaline water: when phenol 0.5 M is added of to this system can observe that a complete passivation of the electrode results after anodic polarization, up to very positive potentials.

When, according to a patent granted to Grace company[32], the system was reproduced on an higher scale using Fe sheets anodes, the formation of thin polyoxyphenylene coatings was verified: however they looked spongy to powdery and their formation required 80–100 Faraday for each mole of phenol polymerized.

The anodic polymerization in situ of substituted phenols was performed with success in alcohols with addition of alkaline bases by Bruno et al.[33]. They found that passivating protective polyoxyphenylene films can be formed on several metal substrates by a fast electrolysis. These films have a thickness ranging from ≈ 0.05 to 0.15 μm, are homogeneous and adherent, their formation in situ being favored by the insolubility in the alcoholic medium. Table 7 illustrates the multistep mechanism suggested for the process which is practically the electrochemical analogue of chemical phenol polymerization catalyzed by metal compounds of higher oxidation number[31]. Although such coatings are very interesting, their extreme thinness give a very poor mechanical protection, thus limiting the practical applications.

The very fast passivation of a metal electrode which results on anodizing phenols in suitable media may be explained in this way: in a medium containing a base a large number of phenol molecules are deprotonated to phenoxide anions which are stored on the positive electrode. Phenol molecules furthermore, owing to the conjugated π-systems which may interact with the ions of a metallic lattice, are likely to adsorb

Table 7. Electrochemical polymerization mechanism of phenols[30]

a) Initiation :

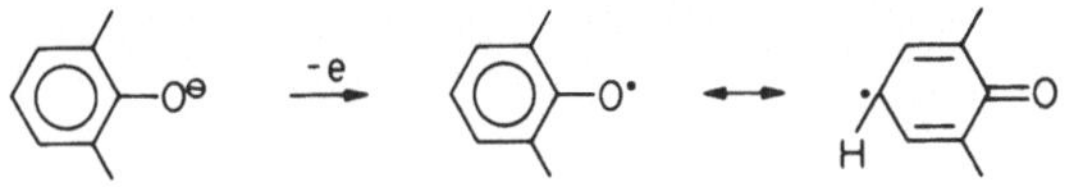

b) Propagation :

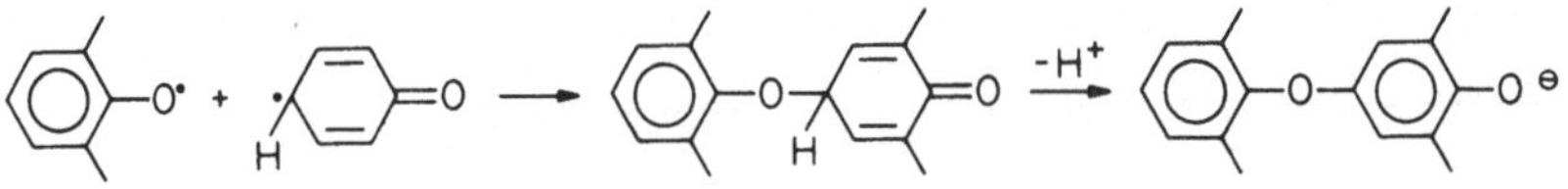

c) Termination :

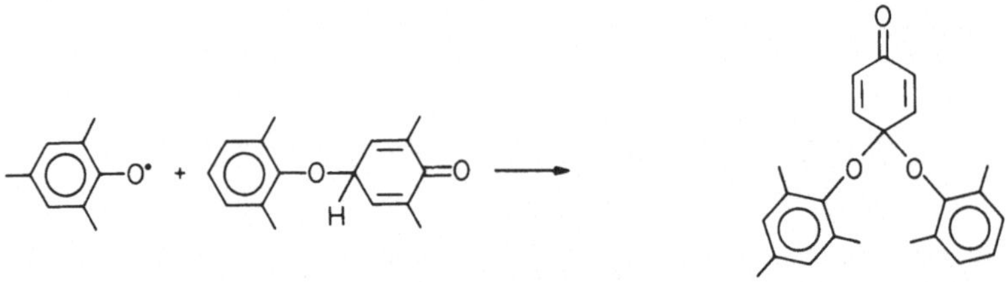

on the anode by assuming a planar position[22]. Thus a fast head to tail polymerization along the electrode plane is particularly favored with the consequent formation of an insulating coating. This picture according to Mengoli et al.[34] is substantiated by the following fact: when a phenol solution in water is salified with an organic base (amine) which may adsorb on the anode, the electropolymerization no longer takes place in situ but develops into the solution as the competitive adsorption of the amine breaks the coupling reaction along the plane of the electrode.

There is, however, a more advantageous intermediate behavior: as a matter of fact Mengoli et al.[34] have devised some compositions based on phenol which do not cause fast passivation of the anode but allow thick polyoxyphenylene coatings to be attained.

A typical composition comprises phenol as the main monomer, minor amounts of a substituted phenol (generally o-chlorophenol) as the auxiliary monomer and ethylenediamine: the solvent basically is water to which some per cent of methanol is added. On varying the ratio between phenol and ethylenediamine and the potential difference applied, finished coatings (that is causing the insulation of the anode) with a thickness ranging from some few μm to ≈ 25 μm could be obtained.

It was found that ethylenediamine is the clue of such realization as results from the following example:

A phenol/o-chlorophenol (mole ratio 10/1) solution buffered at pH = 10.5 with NaOH was electrolyzed at Fe anodes by various potential differences applied during different times: the resulting polyoxyphenylene coatings at the best were below 0.05 mg/cm^2. Conversely when to the same solution 1% ethylenediamine was added, under the same conditions of electrolysis homogeneous, good coatings reaching

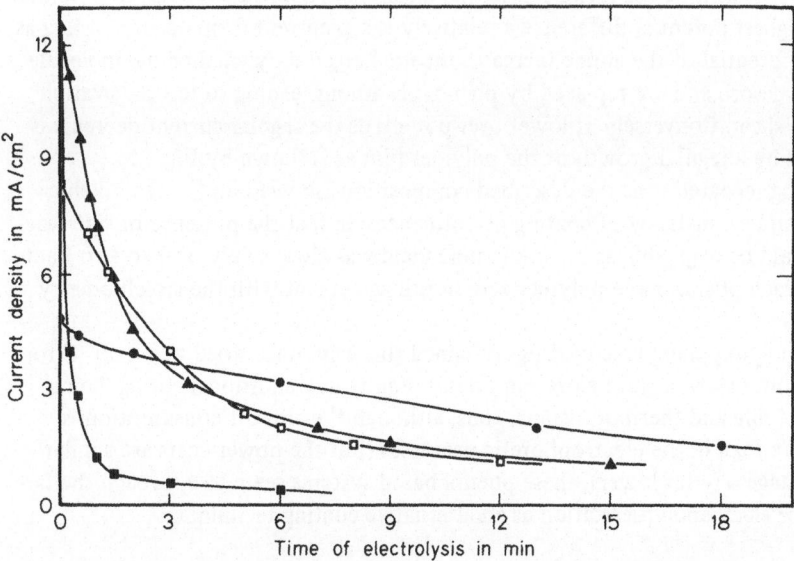

Fig. 22. Passivation rate of a Fe sheet anode during the electrolysis of phenol (1.42 M)/o-chlorophenol (0.16 M)/ethylenediamine (0.45 M) in H_2O/CH_3OH for various voltages: ■ 7.0 V; ▲ 5.5 V; □ 4.5 V; ● 3 V

1 mg/cm^2 in weight were obtained: further additions of ethylenediamine improved the yields still further.

Ethylenediamine not only affects the kinetics of phenol polymerization but it enters also into the coating composition: elemental analysis in fact could show the presence of amine in the polymer film with concentrations up to ten times higher than in the electrolysed solution. This fact also represents a test of a likely ethylenediamine storage at the anode surface: a further indirect test of its adsorption on Fe is given by Fig. 22 which shows, for a typical composition, the variation of current

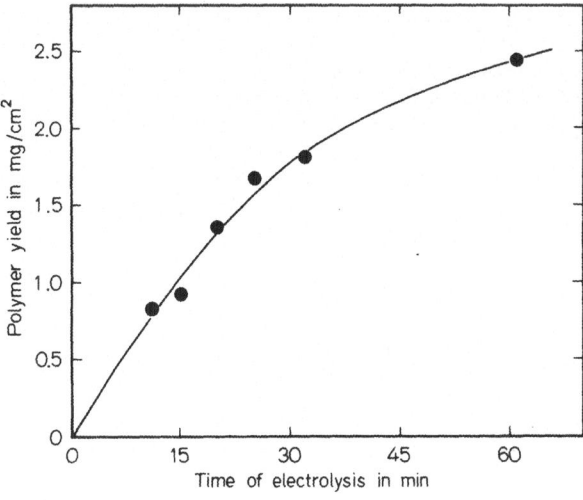

Fig. 23. Polymer yield obtained from the system of Fig. 22 with 3 V applied as a function of the electrolysis time

with electrolysis time for various potential differences applied. One has to note that for the highest potential difference a relatively sharp current drop occurs: in fact as the over-potential of the anode increases the uncharged ethylenediamine molecules probably desorb and are replaced by phenoxide anions leading to fast passivation of the Fe sheet. Conversely at lower over-potentials the regular current decrease is paralleled by a regular growth of the polymer film as is shown by Fig. 23.

During repeated runs the described compositions showed both a remarkable stability and a constancy of coating performances; in fact the presence of side reactions should be negligible as the coulombic yield was always very near to two Faradays for each phenol mole polymerized, in full agreement with the stoichiometry of Table 7.

The polyoxyphenylene coatings obtained this way, once dried at $\approx 100\,^{\circ}C$ for a few minutes show a good gloss and do not require any additional curing, being quite insoluble and thermoresistant: thus, although the current consumption is higher here than in the electrophoretic processes (but the power costs are similar as the voltages are far lower), these phenol based systems seem to approach the targets of the electropolymerization as an alternative coating technique.

7 References

1. Yeats, R. L.: Electro-painting. Teddington: Robert Draper 1970
2. Cooke, B. A., Ness, N. B., Palluel, A. L. L.: The electrodeposition of paint. In: Industrial electrochemical processes. Kuhn, A. T. (ed.) p. 417. Amsterdam—London—New York: Elsevier 1971
3. Brewer, G. E. F.: Electrodeposition of coatings, Advances in chemistry series 119, Washington, D. C.: Am. Chem. Soc. 1973
4. Funt, B. L., Tanner, J.: Electrochemical synthesis of polymers. In: Weinberg, N. L., (ed.): Technique of electroorganic synthesis, Vol. II, p. 559. London: John Wiley & Sons 1975
5. Mengoli, G., Tidswell, B. M.: Polymer *16*, 64 (1975)
6. Asahara, T., Seno, M., Tsuchiya, M.: Kinzoku Hyomen Gijutsu *20*, 64 (1969); idem ibid. *20*, 411 (1969); idem ibid. *20*, 617 (1969)
7. Kaneko, H: Kinzoku Hyomen Gijutsu *21*, 93 (1970)
8. Bezuglji, V. D., Korshikov, L. A., Kravtsova, L. I., Bandarenko, I. B., Fioshin, M. Ya.: Elektrokhimiya *8*, 1658 (1972)
9. Korshikov, L. A., Kravtsova, L. I., Bezuglyi, V. D.: Elektrokhimiya *10*, 106 (1974)
10. Korshikov, L. A., Karpinetes, A. P., Bezuglyi, V. D.: Elektrokhimiya *10*, 990 (1974)
11. Bockris, J. O'M., Reddy, A. K. N.: Modern electrochemistry, Vol. II. New York: Plenum Press 1970
12. Levin, E. S., Yamshchikov, A. V.: Elektrochemical reduction mechanism in organic peroxides. In: Frumkin, A. M., Ershler, A. B., (eds.): "Progress in electrochemistry of organic compounds", p. 319. London: Plenum Press 1971
13. Barnes, D., Zuman, P.: Trans. Faraday Soc. *65*, 1668 (1969)
14. Grace, W. R. et al.: Br. Pat. 1.134.387 (1967), Chem. Abstr. *70*, 30197 g; idem, Br. Pat. 1.179.543 (1969), Chem. Abstr. *72*, 101933 g
15. Jostan, J. L., Krusemak, W., Bogenschütz, A. F, Oberfläche-Surface *10*, 677 (1969)
16. Bogenschütz, A. F., Jostan, J. L., Krusemark, W.: Galvanotechnik *60*, 750 (1969)
17. Oliver, J.: Prod. Finish (Cincinnati) *34*, 60 (1970)
18. Sobiesky, J. F., Zerner, M. C.: U.S. Pat. 3.464.960 (1969), Chem. Abstr. *71*, 103294 n
19. Collins, G. L., Thomas, N. W.: J. Polymer Sci., Polymer Chem. Ed. *15*, 1819 (1977)

20. Garg, B. K., Raff, R. A. V., Subramanian, R. V.: U.S. NTIS, AD Rep. 1977 from Gov. Rep. Announce. Index (U.S.) 1977 (7711) 158; Chem. Abstr. *87*, 5458614
21. Teng, F. S., Makalingam, R., Subramanian, R. V., Raff, R. A. V.: J. Electrochem. Soc. *124*, 995 (1977)
22. Damaskin, B., B., Petrii, O. A., Batrakov, U. V.: Adsorption of organic compounds on electrodes. New York: Plenum Press, 1971
23. Hodes, H. A., Sobieski, J. F., Zerner, M. C.: U. S. Pat. 3.554.882 (1971); Chem. Abstr. *74*, 65723K
24. Kolthoff, I. M., Ferstanding, L. L.: J. Polym. Sci. *5*, 563 (1951)
25. Hodes, H. A., Sobieski, F., Zerner, M. C.: U. S. Pat. 3.650.909 (1972); Chem. Abstr. *77*, 7471 c
26. Grace, W. R. et al.: Br. Pat. 1.155.497 (1969); Chem. Abstr. *71*, 41328 m
27. Fr. Pat. 2.193.067 (1973); Chem. Abstr. *80*, 147037C
28. Shimizu, S., Shinohara, H., Kondon, T., Yamada, N.: U.S. Pat. 3.865.617 (1975); Chem. Abstr. *78*, 161041 q
29. Mengoli, G., Daolio, S., Giulio, U, Folonari, C.: J. Appl. Polym. Sci., in press
30. Mengoli, G. et al. to be published
31. Hay, A. S.: Adv. Polym. Sci. *4*, 496 (1967)
32. Grace, W. R. et al.: Br. Pat. 1.156.309 (1969); Chem. Abstr. *71*, 72049
33. Bruno, F., Pham, M. C., Dubois, J. E.: Electrochim. Acta *22*, 451 (1977)
34. Mengoli, S., Daolio, S., Giulio, U., Folonari, C.: J. Appl. Electrochem., in press

Received December 15, 1978
W. Kern (editor)

Electroinitiated Polymerization on Electrodes

R. V. Subramanian

Department of Materials Science and Engineering, Washington State University, Pullman, WA 99164

Electroinitiated polymerization reactions leading to the formation of polymer on electrode surfaces – as distinguished from that in the bulk of the electrolytic solution – are discussed here. The structure and morphology of the polymer, and its bonding and adhesion to the electrode surface are examined in addition to the mechanisms of polymerization. The formation of adherent coatings on electrodes discussed in this manner is that of polymers from not only vinyl monomers, but also from phenols, acrolein, benzonitrile, polyimide intermediates and phenylacetylene. The novel use of carbon fiber electrodes in electroinitiated polymerization is introduced and its significance to interphase modification in carbon fiber composites is brought out. With continuing advances in techniques for the study of the formation, control, and properties of polymer films on electrodes, concurrent progress in investigation of surface coatings and graphite composites is indicated.

Table of Contents

I Introduction

With the demonstration of the tremendous potentialities of electroorganic chemistry, the application of electrochemical techniques to polymerization reactions has received increasing attention in recent years. A number of reviews during the last decade, the most recent being that by Funt and Tanner in 1975, have served to summarize studies on electroinitiated polymerization of monomers[1-4]. In these reviews, attention has been focused on electrochemical polymerizations in solution in which the propagation reaction takes place in the bulk of the solution though the initiating species are formed by heterogeneous electron transfer reactions occuring at the electrodes. Unlike in such studies where the coating of electrodes, whenever it occurred, had been an unexpected or undesirable side reaction since it changes the electrode potential in an unquantifiable manner making the control of the process or study of the reaction mechanism difficult, the objective of a current investigation in the author's laboratories is to explore the use of electropolymerization to coat electrode surfaces with polymers formed by electroinitiated polymerization *in situ*[5-10]. This should be distinguished from conventional electrocoating methods where a preformed ionizable polymer is deposited on substrate from solutions or suspension of polymer by electrophoretic means.

The formation of adherent films of poly(*p*-xylylenes) on aluminium cathodes by electrolysis of a solution of *p*-xylylenebis(trimethylammonium salt) in a polar solvent is reported in an early paper by Ross and Kelly[11]. The formation of polymer films on metal electrodes by electropolymerization of a number of vinyl monomers has been shown in exploratory studies by Asahara and co-workers[12-18]. More systematic studies of electropolymerization on electrodes are few and very recent in origin. The results of these studies as well as those of our investigation are discussed in this paper. "Feasibility of polymer film coatings through electroinitated polymerization in water medium" is the subject of a companion paper in this journal by G. Mengoli. This aspect of the subject, therefore, will not be emphasized in the present review.

A novel adjunct to this research is our investigation of polymer coating of graphite fibers by electroinitiated polymerization, and of the effect of such coating on the properties of composites prepared from the coated graphite fibers[6, 7]. The research has also led to the electropolymerization of some unusual monomer systems, notably the electroinitiated polymerization through C≡C and C≡N bonds[10]. These results are also discussed in some detail in this paper.

II Electropolymerization on Electrodes

a) Vinyl Monomers

Most of the vinyl monomers have been electropolymerized on metal electrodes though acrylonitrile has been the monomer given particular attention. As mentioned earlier, Asahara, et al.[12-18], have explored the formation of polymer coatings on steel and other metal electrodes from a wide range of monomers in the presence of various supporting electrolytes. While their experiments scanned a range of experimental parameters, an interesting observation was that of the formation of "whiskers" during the electrolytic polymerization of acrylonitrile onto an aluminium metal surface in the presence of acetonitrile, styrene, toluene, xylenes, or benzene as solvent. X-ray analysis indicated that the "whiskers" were amorphous polyacrylonitrile of molecular weight, about 1000. The phenomenon of "whisker" formation from acrylonitrile, confirmed by other workers, is discussed later in more detail.

Bezuglyi, Korshikov and co-workers[19-21] also obtained poorly adhering poly-(methyl methacrylate) and polystyrene films on steel cathodes by electrolysis of the solutions of corresponding monomers from N,N-dimethylformamide (DMF). Through the use of free radical inhibitors and anionic inhibitors, the mechanism of polymerization was established to be anionic. This mechanism of electropolymerization was confirmed, as in other studies of polymerization, by copolymerization of styrene with methyl methacrylate under identical conditions[22,23]. Interestingly, the authors have examined the yield and composition of the copolymers separately on the electrode surface and in solution and seem to find evidence for an anionic mechanism on the surface and a combined anionic and free radical mechanism in the DMF solution containing Bu_4NClO_4 as supporting electrolyte[23].

Suitable conditions for the formation of polymeric coatings by electrochemical polymerization have been investigated by Mengoli and Tidswell who studied the cathodic polymerization on steel of several acrylic and methacrylic esters in DMF or dimethyl sulfoxide (DMSO)[24]. From an examination of the data it was rightly emphasized that in aprotic media the monomer/solvent couple has to be selected such that the solubility parameter of the polymer is sufficiently different from that of the solvent, or that at least the time scale of building up the film should be lower than the time required for the dissolution of the polymer. From the effect of substituent size and shape on the type of coating obtained, indirect indications were also obtained that the adsorption of monomer molecules on the electrode before polymerization is very important for the subsequent process in situ to occur. These indications find confirmation in related observations in other investigations[5,6]. In copolymerizations of methacrylonitrile with isobutyl methacrylate, the variation of copolymer composition with monomer feed ratio was almost that of an alternating copolymer. However, the typical coloration of conjugated carbon-nitrogen bonds shown by pure polymethacrylonitrile was present also in the copolymer containing less than 50 mole % methacrylonitrile. This was taken to indicate the initial formation of blocks of methacrylonitrile in the polymer at the electrode surface into which isobutyl methacrylate then diffuses faster.

The use of a crosslinking monomer to produce coherent coatings of otherwise soluble polymer is illustrated by the acrylamide/N,N'-methylenebisacrylamide system. In the presence of $ZnCl_2$, aqueous solutions of these monomers are found to be electropolymerized on metal cathodes. It hat been proposed that the initiation of polymerization here was by hydroperoxy radicals arising from a series of reactions beginning with the formation of ZnO_2^+ from Zn^{2+} and O_2[25,26]. In a recent reexamination of the kinetics of polymerization in this system, evidence has been obtained that the species involved is $ZnOH^+$ which is complexed with the monomer by multi-site attachment in a ring structure[27]. The pH dependence of electrocoating, the observed electropolymerization of N,N-dimethylacrylamide and the correlation of electropolymerization with carbonyl shifts detected in laser Raman spectra of a selected series of acrylic monomers are advanced as evidence for these conclusions.

The electropolymerization of diacetoneacrylamide[1] (DAA) has been successfully employed for the formation of polymer coatings in a number of patents[28-30]. The polymerization, generally conducted in acidic solutions, is probably initiated by H atoms produced by the discharge of protons at the cathode[31]. Bogenschütz and co-workers have studied in detail the cathodic polymerization of DAA on a number of metal electrodes and have also examined the nature of the polymer coating formed[31-34]. Under identical conditions, the thickness of the poly(diacetoneacrylamide) (PDAA) coatings formed increased with increase in hydrogen overvoltage of the metal electrodes studied which included Sn, Pb, Zn, Cd, Zr, Al, Ag, Cu, Fe, Ni, Pt, Mo, W, Au, and also porous and graphitic carbon.

Polymerization was initiated after a short induction period and ceased after the build-up of a film of maximum thickness on the metal cathode which was accompanied by a drop in current density. Metal substrates having large hydrogen overvoltages were coated more easily than those on which hydrogen was discharged easily, probably because the adsorbed hydrogen on the electrode surface obstructed the formation of polymer. Thus, the more electronegative metals were easier to coat than those which were less negative. Also, porous surfaces could be coated more quickly than smooth ones, apparently because of the greater surface reactivity in the former case. The electropolymerized coatings seemed to be superior in some respects like dielectric loss when compared to the usual polymer coating formed by other methods; the breakdown voltage of electropolymerized PDAA was remarkably high, up to 300 kV/cm at only a few millimicrons thickness. The dielectric constant of electrochemically polymerized PDAA lay between 2.5 and 2.8 (at 800 Hz) while that of the polymer film prepared by thermal initiation was between 2.1 and 2.5.

The crystallizability of the polymers also showed significant differences[32]. The electroinitiated polymer showed a diffuse halo in X-ray diffraction initially which sharpened to well defined rings after the polymer was annealed above its melting point, at about 200 °C. The crystallizability of the polymer, as revealed in these experiments, suggests a stereospecific control of polymerization on the electrode surfaces. The ir spectra of the polymer confirmed that polymerization had indeed taken place through the vinyl groups.

The factors that control the coating thickness, the morphology of the polymer deposit and the adhesion of the polymer formed on the metal substrate, have been studied by us for a number of monomers in both aqueous and non-aqueous solvent media[5]. Both cathodic and anodic polymerizations and copolymerizations were utilized to form coatings on metals from vinyl monomers as well as by ring opening reactions of cyclic monomers. Acrylic acid, acrylonitrile, N-methylolacrylamide (N-hydroxymethylacrylamide), methyl methacrylate, styrene, glycidyl methacrylate (2,3-epoxypropyl methacrylate), phenyl glycidyl ether (2,3-epoxypropyl phenyl

---

1   Systematic nomenclature: N-(1,1-dimethyl-3-oxobutyl)acrylamide.

ether), maleic anhydride, etc. were included in this study of forming coatings on aluminium, copper and steel cathodes. It was first established that the electric field obtained under the experimental conditions employed did not cause the formation of polymer coating, since, in the absence of added electrolyte, neat monomers such as styrene or acrylonitrile caused high cell resistances, negligible or no electric current (i.e., no electrode reaction) and no polymer formation. However, monomers, vinyl as well as cyclic, which can be polymerized by free-radical or ionic mechanisms could be polymerized onto metal electrodes by electrolysis of their solutions in stuitable solvents containing a supporting electrolyte. The polymerizations were conducted in the middle chamber, containing monomer, of a three compartment electrolytic cell at constant dc voltage.

Based upon morphology as determined by light and electron microscopic techniques, the superficially uniform polymer coatings obtained could be subdivided into three classes. The class-I coatings had powdery to spongy appearance and could be made to grow to several hundred micrometers in thickness. The class-II coatings had the appearance of a painted-on coating and generally could be made to grow to 100 μm in thickness. These coatings, when removed from the cell, were always highly swollen with solvent. The class-III coatings were extremely thin, and, consequently, their presence on metal electrode was most difficult to confirm. These coatings could be made to grow to 1 μm in thickness by continuing electrolysis for longer periods of time.

There was a predominant tendency to form powdery to spongy class-I coatings. This tendency is attributed to the polycrystallinity of metal substrate and numerous microscopic irregularities on it, which led to nonuniform tendencies of electrons to escape from the surface as well as provide nuclei for growth of polymer particles. Further, it was found that an increase in the solubility of polymer, formed by electrolysis, in the electrolytic solution counteracted the formation of powdery to spongy coatings and led to the formation of a coating that had a "painted on" appearance. Good solubility of polymer formed in situ in the electrolytic solution caused extremely thin-class-III coatings to be formed.

The current drop observed during formation of the coatings was much smaller than would be expected on the basis of calculated resistivity of uniform polymer coatings. This could have been caused by the swelling of coating by solvent in some cases and also by the entrapment of solvent in the porous coatings.

For class-II PAN coating, formed from AN-DMF-NaNO$_3$, the dependence of coating thickness on cell voltage, initial current, monomer concentration and time of electrolysis was determined. The results indicated that chain polymerization kinetics were followed to a considerable degree. The cell voltage, above a critical value, had negligible effect on coating thickness, in spite of the fact that higher cell voltage does lead to substantially increased current. Perhaps higher cell voltage causes increased charge transfer to species other than monomers which do not initiate polymerization. It is also likely that when large concentrations of initiating radicals are formed at high current, they fail to grow to high polymers and suffer early termination to form soluble low molecular weight products. The coating thickness increased linearly with increasing monomer concentration, except at lower concentrations of AN.

The formation of "whiskers" as reported by Asahara, et al., [16)] can be observed in the above system also. There is a minimum critical thickness depending on the conditions of polymerization below which whisker formation does not take place. If the coating is allowed to grow beyond this thickness, it generally peels, with the peeled sections aligning themselves in the direction of the electric field. The peeled sections do not, however, detach themselves completely from the electrode's surface. Thus, they give the appearance of whisker formation.

An example of such whiskers is shown in Figure 1. This phenomenon was unique to electropolymerization of AN monomer in DMF and was not observed during electropolymerization of either methyl methacrylate or styrene under identical conditions. It could therefore be inferred that the participation of nitrile groups in the polymerization reaction was responsible for whisker formation. Thus, during the initial stages of electrolysis, a relatively fast vinyl polymerization leads to formation of polyacrylonitrile coating on the cathode. Subsequently, some of the nitrile groups

$$
\begin{array}{cccc}
\diagdown CH_2 \diagup \diagdown CH_2 \diagup \diagdown CH_2 \diagup \diagdown CH_2 \diagup \\
CH CH CH CH \\
| | | | \\
C C C C \\
\diagdown N \diagdown N \diagdown N \diagdown N
\end{array}
$$

accept an electron to become anion-radicals $-\dot{C}=N^-$ thus initiating anionic polymerization of nitrile groups leading to cyclized polyacrylonitrile shown above. If the anionic polymerization of nitrile groups were slow compared to vinyl polymerization, a build-up would result of anions in the polyacrylonitrile coating on the cathode.

Fig. 1. Polyacrylonitrile "whiskers" in electropolymerization of acrylonitrile (NaNO$_3$/DMF)

This is so because such anions are covalently bonded to the precipitated polymer, and thus, cannot move away from the cathode. It is apparent that as the concentration of anions in the coating increases above some critical value, the coulombic repulsion between the cathode and the negatively charged coating will be sufficient to cause the peeling of the coating from the cathode surface. Such peeling is indeed observed and leads to formation of so called "whiskers" as shown in Fig. 1. Since the peeled portions of coating still contain covalently bonded anions generated from nitrile groups, these portions align themselves in the direction of the electric field as was also observed.

The structural features of PAN as revealed by its ir spectrum and also its characteristic orange coloration, lent further support to the formation of the cyclized PAN structure shown above. In polymer films obtained by electrolysis of the acrylonitrile/sodium nitrate/DMF system, existence of cyclized polyacrylonitrile was shown by absorption at 1570 cm^{-1} for conjugated $-C=N-$ bonds[35, 36], at 1650 and 1635 cm^{-1} for C=N bonds, and at 1220 and 1145 cm^{-1} for C$-$N bonds[37, 38]. Extraction of the coating with DMF at 153 °C gave a residue of 48% for 48 h of extraction and 9% for 72 h of extraction. Further, when the extraction was carried out at room temperature, a residue of 40% was obtained after two weeks of extraction. This shows that the solubility of the product obtained is much less than the solubility of polyacrylonitrile in DMF. Using DMF at 35 °C as solvent, the molecular weight of the soluble portions was determined to be 20000 in each case by solution viscosity method. For this purpose, the parameters of the viscosity-molecular weight relationship for cyclized polyacrylonitrile were assumed to be the same as for linear polyacrylonitrile[39]. The elemental analysis of the coating gave a C : H : N ratio of 3.01 : 3.03 : 1.00 against expected values of 3.00 : 3.00 : 1.00. Also, the orange-yellow coating obtained continued to darken for a long time after removal from the electrolytic cell. Final coloration was much deeper when drying was carried out at 100 °C than when it was carried out at room temperature. It should be noted that the *thermal* cyclization of polyacrylonitrile does not begin at least until 250 °C[39]. In view of these data, it is concluded that a cyclized polyacrylonitrile coating forms on the cathode when acrylonitrile/sodium nitrate/DMF solution is electrolyzed.

Similarly, the infrared spectra of polymer formed from electrolysis of acrylonitrile/water/sulfuric acid indicated the presence of amino groups. The polymer formed from electrolysis of acrylic acid/water/sulfuric acid, on aluminium cathode, showed the presence of the aluminium salt of acrylic acid.

Next, one must consider the fact that thick polyacrylonitrile and polystyrene coatings were formed from electrolysis of the solutions of corresponding monomers in DMF even though DMF is a good solvent for both polymers, while only thin poly(acrylic acid) coatings could be obtained from electrolysis of its monomer from aqueous solutions. The fact that the coatings form at all during electrolysis of these monomer-solvent combinations indicates that the polymerization in the desolvated layer of monomer near the electrode is strongly favored over solution polymerization under conditions chosen in the present study. It is also suggested that the maximum thickness of coating obtained is a balance between the rate of polymerization and the rate of dissolution of the polymer deposit formed. The rate of anionic polymerization, which predominates in electropolymerizations carried out in DMF, is several orders of magnitude larger than the rate of free-radical polymerization which is prevalent during electropolymerizations from aqueous solutions. When the rate of dissolution is comparable in monomer-solvent combinations, formation of thicker coatings would be favored when an aprotic solvent is used, even if the polymer formed is solubel in the solvent.

The maximum obtainable thickness of coating over a long period of electrolysis at constant cell voltage was found to decrease with increasing solubility of polymer deposit in electrolytic solution. Thus, when the acrylic acid/water/sulfuric acid system is electrolyzed, only thick class-III coatings could be obtained. However, when a water-soluble crosslinking agent such as N,N'-methylenebisacrylamide or aluminium chloride was added to the electrolytic solution to insolubilize the polymer, thick class-I coatings were obtained. In such coatings, the ultimate coating thickness increased with increasing cell voltage.

The formation of rings in the polymer chain by the polymerization of nitrile groups has also been observed in the bulk polymerization of acrylonitrile by Trifonov and Shopov[36]. In this study of the electroinitiated polymerization of acrylonitrile with tetraethylammonium perchlorate as supporting electrolyte, the polymer was formed on a germanium prism which was part of an internal reflection spectroscopy device. The appearance of the absorption band at $1570\ cm^{-1}$ in the spectrum of the yellow colored polymer indicated the presence of conjugated multiple bonds which was also supported by uv absorption at a longer wavelength 304 nm, compared to 271 nm for white polyacrylonitrile prepared by free-radical initiation.

b) Benzonitrile

The occurrence of cyclized polyacrylonitrile during electropolymerization of acrylonitrile suggested that the electropolymerization method might be developed into a procedure for polymerization through nitrile groups. Hence, the electropolymerization of benzonitrile to the corresponding linear, $-C=N-$ conjugated polybenzonitrile $+(C_6H_5)C=N+_n$, was investigated[10]. Using a three compartment electrolytic cell partitioned by fritted glass discs, polybenzonitrile coatings were obtained on various types of cathode materials. The complete results are summarized in Table 1. All polymerization reactions were conducted at room temperature. Typically, a dry solution of benzonitrile in DMF containing lithium nitrate was electrolyzed for up to 72 hours. The loose polymer deposit formed on the cathode was scraped off, thoroughly washed with water and dried.

The presence of small amounts of water completely inhibited the electropolymerization reaction. However, the addition of small quantities of a free radical inhibitor had little effect on polymerization. Hence, it was concluded that the polymerization mechanism is anionic. Further, the reaction rate was rather slow, presumably because of the resonance stabilization of the anions at the ends of growing polymer chains. Such resonance stabilization may also be a limiting factor in the maximum molecular weight achievable by electropolymerization. The intrinsic viscosity of the polymer, measured at 35 °C in concentrated sulfuric acid solution was found to be 0.07 dl/g. The polymer was partially soluble in several organic solvents including

Table 1. Electroinitiated polymerization of benzonitrile

Composition of catholyte[a]	Cathode	Cell voltage in V	Current in mA Initial	Current in mA Final	$\cdot[\eta]^{b)}$ dl/g
1. Benzonitrile (200 g) LiNO$_3$ (5 g)	Steel	12	104	28	0.062
2. Same as above	Zinc	12	103	18	0.073
3. Same as above	Copper	12	117	20	0.070
4. Benzonitrile (80 g) LiNO$_3$ (2 g)	Aluminium	12	64	17	0.068
5. Benzonitrile (80 g) LiNO$_3$ (2 g) H$_2$O (5 g)	Aluminium	12	108	101	No polymer
6. Benzonitrile neat	Platinum	60	0.1	0.02	No polymer
7. Benzonitrile (200 g) (C$_2$H$_5$)$_4$NBr	Zinc	12	74	56	0.062

a   In DMF solution, except 6.
b   Intrinsic viscosity determined in conc. H$_2$SO$_4$ at 35 °C.

DMF but was completely soluble only in concentrated sulfuric acid. From Table 1, it is apparent that the electrode material has negligible influence on the molecular weight of the polymer obtained.

A mass spectroscopic study of the polymer scraped from the metal surface was conducted at low temperature (50–200 °C) in order to volatilize only low molecular weight compounds while simultaneously preventing the thermal degradation of the polymer. The results indicated that the polymer was not predominantly the cyclic trimer 2,4,6-triphenyl-1,3,5-triazine (mol.wt. 309) which is a typical end product of attempts to polymerize benzonitrile at <150 °C by chemical means. The mass spectrum did not indicate a significant excess of mass 309. The sample, however, showed a significant excess of mass 103--presumably the residual benzonitrile, and also of mass 120--presumably benzoic acid and benzamide generated by hydrolysis of the polybenzonitrile during aqueous washing of the sample. Further, the infrared spectrum of the polymer, shown in Fig. 2, indicated the presence of absorption bands for conjugated –C=N– bonds at 1605 cm^{-1} and 1520 cm^{-1} in agreement with published data on polybenzotrile[40].

It should be noted that the convenient electropolymerization of benzonitrile under ambient conditions to a linear conjugated polymer is in striking contrast to the rather severe conditions of temperature and pressure necessary to cause homopolymerization of nitriles by ordinary chemical methods. The preliminary ordering of C≡N dipoles under the influence of an electric field in the electrode region might be a contributing factor in favoring the convenient electropolymerization of nitriles.

Kabanov, et al., [40] have suggested that, based on thermodynamic considerations, the polymerization of C≡N to the corresponding polyconjugated polymers cannot occur under conditions typical for polymerization of vinyl or acetylenic compounds.

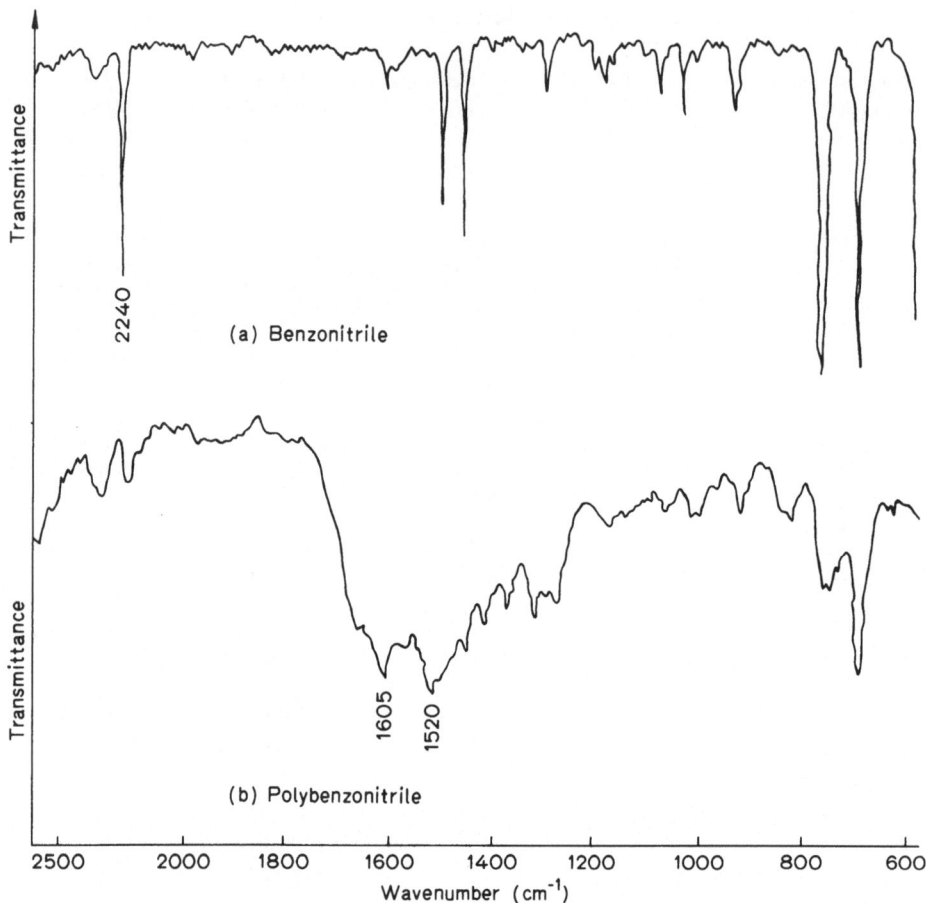

Fig. 2. IR spectra of (a) benzonitrile, using a drop between KCl crystals and (b) polybenzonitrile

Their suggestion was based on the fact that the estimated heat of polymerization at the C≡N bond cannot be substantially more than zero. Since the entropy of polymerization of C≡N type compounds to the corresponding $\{C=N\}_n$ polymers is negative, the reaction will lead to an increase in the free energy of the system, making such a polymerization thermodynamically unfavorable. Kabanov, et al.[40, 41] were able to prepare polyconjugated polymers from acetonitrile, propionitrile, capronitrile, and benzonitrile by first preparing stoichiometric complexes of these monomers with titanium tetrachloride, zinc chloride, or boron trifluoride and subsequently heating these complexes, in the presence of polymerization initiators such as phosphoric acid, to 150–250 °C in sealed ampules. The presence of the complexing agent shifts the chemical equilibrium favorably in the direction of polymerization reaction through a preliminary ordering of the monomer molecules. Liepins, et al.[42, 43] were also able to synthesize polyconjugated polymers from nitriles. They used free radical initiators at temperatures of 150–330 °C.

The polyconjugated polymers derived from nitriles have been shown to have useful semiconducting properties[44-46]. Similarly, the trimerization of nitrile groups has
been used as a curing device, similar in purpose to epoxy curing reactions, to overcome processing difficulties associated with high temperature resistant aromatic poly-
imides and other thermally stable polymers[47-49]. From the above discussion, it is
obvious that a convenient method of converting $C\equiv N$ compounds to corresponding
$+C=N+_n$ polymers as by electropolymerization, would have many practical applications. The mechanism of this polymerization requires further investigation.

c) Phenols

The coating of metals by poly(arylene oxides) formed by electrolytic polymerization
of phenols in aqueous alkaline solution is described in a patent issued to Grace and
Co.[50]. The coated metals, Fe, Ni, Cu, Ti, Pd, etc., were claimed to have improved
abrasion, corrosion and oxidation resistance. Similar pinhole-free, thermally stable,
acid resistant coatings of crosslinked polymers are formed by electropolymerization
of phenol or *ortho*-cresols in bulk or in polar solvents such as acetonitrile or $C_6H_4Cl_2$
in the presence of a base such as triethylamine[51]. The basic conditions in these experiments are conducive to the removal of an electron from the phenoxide anion to
give a phenoxy free radical. On the other hand, in the absence of a base in acetonitrile the anodic oxidation of 2,6-xylenol on a platinum electrode proceeds first
through an electrophilic attack on the aromatic nucleus of the nonionized phenol
with the irreversible removal of two electrons to give a mesomeric phenoxonium
ion[52]. The main product of the reaction is a dimer:

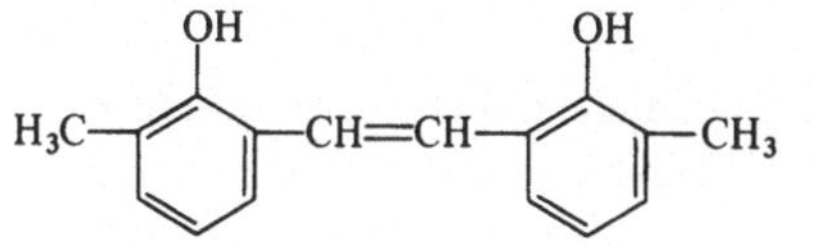

Dubois and co-workers have recently investigated in detail the electrochemical
deposition of poly(phenylene oxide) films on metallic surfaces by electrolysis of
disubstituted phenols under very basic conditions from alcohol solution[53]. The
electrodes were prepared by vacuum deposition of metals on glass plates. The experimental technique involved the use of polaromicrotribometry consisting of measuring the coefficient of dynamic friction of a quartz slider moving at a very slow
speed (8 $\mu m/s$) at the electrode surface. Simultaneous recording of the current/poten-
tial and friction/potential curves allowed great precision in revealing surface modification by electrochemical reaction. Thus, the stick-slip friction curve characteristic
of sliding on metal was changed, at the onset of oxidation of the xylenol, to a steady
line with the simultaneous formation of a film on the metal surface. The nature of
the product film on the metallic mirror was examined by multiple reflection ir spectroscopy. The spectrum identified the polymer as poly(2,6-dimethyl-1,4-phenylene
oxide). The polymerization is suggested to proceed by the initial oxidation of the

phenolate ion and head-to-tail coupling of the resulting phenoxy radicals followed by deprotonation to yield the propagating species

which can again undergo anodic oxidation. Polymerization occurs on any metallic surface provided that the oxidation potential of the metal is greater than the discharge potential of the phenolate ion. Bulky substituents at the 2,6-positions prevent polymerization by hindering the head-to-tail coupling of the phenoxy radicals. In such cases, the biphenoquinone is obtained by tail-to-tail coupling. The position and the nature of the substituents influence the reaction by variations in the stabilization of the phenoxy radical. The poly(phenylene oxide) film has been found to be adhesive, insulating, hydrophobic, and resistant to acids as well as bases. These properties of the surface coating would seem to offer great potential for the application of the electropolymerization technique for the formation of anticorrosive and electrically insulating protective films on clean metal surfaces.

Heat resistant poly(oxyphenylene) coatings have also been prepared by the electrochemical initiation of the polymerization of 2,6-dimethylphenol in the presence of $Cu(OAc)_2$ in admixture with traces of $CuCl$[54]. While the mediation of cupric-cuprous couple in the oxidation reaction can be surmised, the details of the mechanisms of the polymerization are not established.

d) Acrolein

The polymerization of acrolein in bulk by electrochemical oxidation or reduction has been studied earlier by Schulz and Strobel[55,56]. More recently, Desbene-Monvernay, et al.[57], have investigated the deposition of polyacrolein by polymerization of acrolein in N,N-dimethylformamide and acetonitrile using benzyltrimethylammonium perchlorate (BTAP) and tetrabutylammonium perchlorate (TBAP) as supporting electrolytes. Metal electrodes were prepared by the deposition of the films of Pt (sputtering), Au, Al, Cu, Ni, and Fe (vacuum deposition) on glass plates. The polymer formed on the metallized plates was characterized by elemental analysis, multiple reflection ir spectra and ESCA, and found to be made up principally of $\{CH_2-CH\}$ units. The polymerization is thus indicated to occur mainly through
   |
   CHO
the ethylenic bonds. It is suggested that the initiating species, CH_2-CH^- radical
                                                                        |
                                                                       CHO
anions, are formed by the reduction of acrolein, though the formation of other initiating species such as OH^- could not be ruled out. Inhibition of polymerization by water or CO_2, but not by oxygen, was seen to support an anionic mechanism for the reaction. Polaromicrotribometry was employed to measure the friction coef-

ficients of the thin films, while their resistivity was measured by depositing a thin gold layer on the organic film to yield a (Pt/film/Au) sandwich. The adhesiveness of the deposits was dependent on the nature of the solvent and supporting electrolyte, and the electrolysis potential, the most adhering films being formed in (DMF + BTAP). The resistivity of the thin layer sandwiches was about $10^6 \Omega \cdot$ cm, though the frequent short circuiting observed indicated the presence of holes in the organic film. The homogeneity of the film was principally a function of the applied potential.

e) Coulombic Effects in Electroinitiated Copolymerization

It is interesting to note that there have been several studies of electrocopolymerization of monomers forming charge transfer complexes[58-60] because of the advantages associated with low temperature initiation to such polymerizations. However, there have been very few studies of electrocopolymerization as such. Asahara, et al.[15], reported the copolymerization of acrylonitrile with styrene and also with several esters of acrylic and methacrylic acids on steel cathodes. They indicated that the electrode metal was effective only as a source of electrons for the copolymerization reactions. However, Yamazaki, et al.,[61] found that both the electrode material and the solvent strongly influenced the composition of the copolymer obtained by electropolymerization of styrene with methyl methacrylate. The observed differences in composition were explained in terms of the preferential absorption of one of the monomers, as was done by Mengoli and Tidswell[24] discussed earlier.

The research described below was undertaken as an in-depth investigation of the role of coulombic repulsion between an aluminium cathode and an ionized monomer in determining the composition of the resulting copolymer[8]. The copolymerization studied was that of methacrylic acid with N,N'-methylenebisacrylamide. A three compartment electrolytic cell with two fritted disk separators was used. An auxiliary platinum electrode was placed in each of the two end compartments while an aluminium strip, to which a surface treatment based on chromic acid etching had been applied, was placed in the middle compartment. The monomer solution contained methacrylic acid (0.436 M) and N,N'-methylenebisacrylamide (0.145 M) in water. The pH of the above solution was adjusted to desired values by adding either sulfuric acid or aqueous sodium hydroxide. The monomer solution (400 ml) was placed in the middle compartment while two end compartments contained water adjusted to the same pH as the monomer solution. While bubbling nitrogen through the monomers solution, the electrolysis was conducted at 10 volts for 6 hours such that the cathodic reaction would occur in the center compartment. The polymer coating obtained on the aluminium cathode was washed with fresh water, dried in vacuum oven at room temperature, and weighed. The copolymer was then scraped off from aluminium and its composition determined by elemental analysis.

The pH of the solution is known to significantly affect the rates of homopolymerization of acrylic and methacrylic acids[62], and the composition of copolymers of these monomers with N-vinylpyrrolidone[62, 63] or with acrylamide[64] from aqueous solutions. Therefore, to separate the effect of electroinitiation and the effect of pH of solution upon the composition of copolymer, a series of copolymerization experiments was carried out as control under identical conditions but using redox initiation instead of electroinitiation. For this purpose, the 1:1 sodium sulfite-potassium persulfate redox initiator system which operates satisfactorily in both acidic and alkaline aqueous solutions was employed[60]. The quantity of initiator added varied from 0.5% to 4% of monomer moles in solution depending upon the pH of the solution. The copolymer yield was limited to less than 10% by trial and error.

The results of electrocopolymerization and redox initiated copolymerization of methacrylic acid with N,N'-methylenebisacrylamide are presented in Fig. 3[8]. The selection of these two monomers for the present study was based upon the fact that these monomers do not undergo significant side reactions, other than the main vinyl polymerization reaction, during electrolysis of their aqueous solutions[5]. Of these two monomers, only methacrylic acid ionizes in aqueous solutions. It was, therefore, anticipated that the methacrylic acid ion will experience a coulombic repulsion from the aluminium cathode which will affect its incorporation in the copolymer forming on the cathode. Further, by controlling the degree of ionization of methacrylic acid and the potential of aluminium cathode, an electrolytic control of the copolymer composition might become possible.

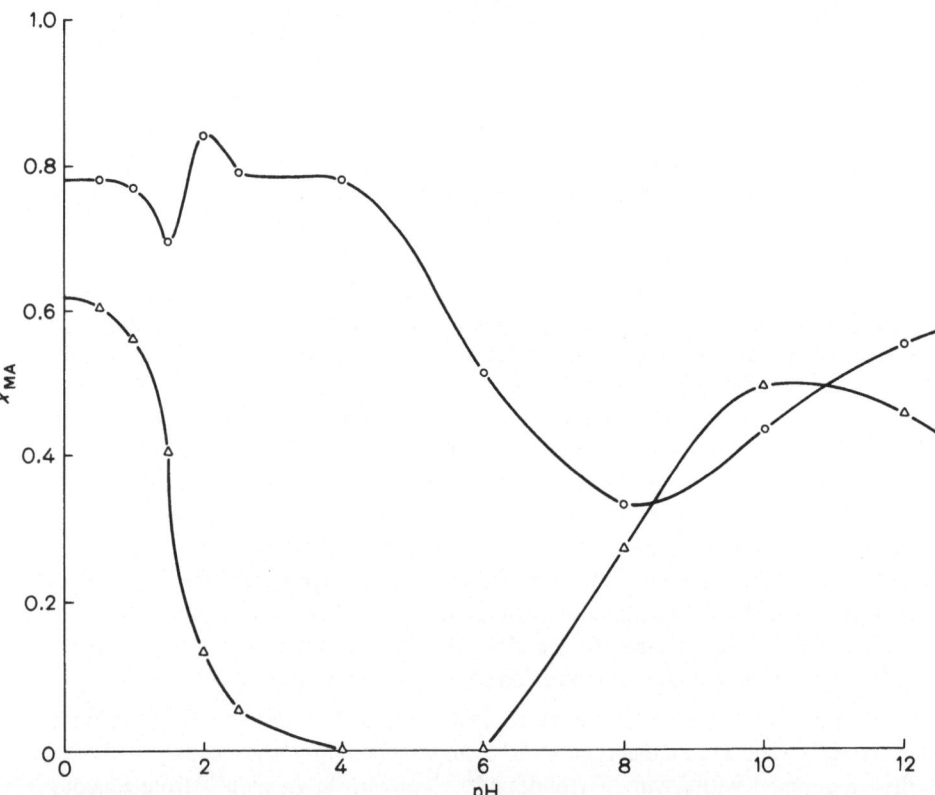

Fig. 3. Mole fraction (X_{MA}) of methacrylic acid in copolymer from methacrylic acid and N,N'-methylenebisacrylamide: (o) redox initiation and (Δ) electroinitiation

In Fig. 3, the mole fraction of methacrylic acid units in the copolymer is plotted as a function of the pH of the monomer solution. The two curves shown compare the composition of the copolymer obtained by the electroinitiation method with that of the composition of copolymer prepared by redox initiation method while keeping the monomer concentration constant and varying the pH of the solution. From this figure, it is clear that the fraction of methacrylic acid units incorporated in the copolymer obtained by electroinitiation on aluminium cathode is generally less than the fraction of methacrylic acid units in the copolymer prepared by redox initiation at all values of solution pH. This difference is particularly large at monomer solution pH from 2.5 to 6.0. This is to be expected since the pKa of methacrylic acid monomer is 4.32 while that of poly(methacrylic acid) is 7.0[62]. The pKa of the copolymer of methacrylic acid with N,N'-methylenebisacrylamide may be assumed to be similar to the pKa of the homopolymer of methacrylic acid. Generally, two mechanisms have been suggested to explain variation of composition with pH for copolymerization of methacrylic acid (or acrylic acid) with other nonionizable vinyl monomers from aqueous media. These are the repulsion of methacrylic acid anion from the anion of growing copolymer chain in low solution pH ranges and the effect of shielding of anions by sodium cations at high solution pH ranges[62-65].

The actions of both of these mechanisms are reinforced when the copolymerization of ionizable vinyl monomers is initiated electrolytically. Thus in low solution pH range, the coulombic repulsion between the methacrylic acid anion and the anion of growing copolymer chain at the aluminium cathode is enhanced by the negative potential of the cathode leading to an even smaller incorporation of methacrylic acid in the electroinitiated copolymer. Similarly, at high solution pH values, the discharge of sodium cations at cathodes leads to even greater shielding of anions of growing copolymer chain, thus, causing an increased addition of methacrylic acid units to the copolymer. It was seen that the dissolution of aluminium cathode in cathodic solution became large at alkaline pH as measured by negative weight gain of the electrode even though a polymer film was always found on the substrate. The role of this dissolved aluminium discharged into the cathodic solution, to the shielding of anions in this solution is not yet clear.

f) Polyimides

Following their work on the electrolytic preparation of polyimide coatings by deposition of polyamic acid precursors onto metal electrodes[66], Phillips and co-workers have investigated the electropolymerization of aminophthalic acids[67, 68]. Before this study, very little work had been done on systems involving electrochemical condensation polymerization which has been defined by Laube and Higgins[69] as polymerization which proceeds with the addition or removal of one or more electrons per electroactive unit of a monomer followed by bond formation among the oxidized or reduced species or unreacted monomer. The latter authors showed that the electrochemical reduction of the carbonyl groups in 4,4'-diacetyldiphenylethane at a mercury cathode in 80% ethanol/1 M potassium acetate yielded polypinacols, primarily oligomers as indicated by low inherent viscosities.

The experiments of Phillips with 4-aminophthalic acid showed that it was electropolymerized in almost quantitative yield to low molecular weight polymer at a platinum anode. Systems involving the acid, amine salt, and the ammonium salt in N,N-dimethylacetamide were explored; the acid and/or the amine salt of the acid provided the best system for polymer formation. Infrared spectral data of reaction components and the higher inherent viscosity of the product in the anode compartment indicated that the polymerization took place in the anode compartment though some migration of intermediate products to the cathode compartment was also observed. Though the exact mechanistic path was not elucidated, it is believed that the process involves initial ionization of the carboxy group to form the carboxylate anion. This species could then be oxidized at the anode to form the cation radical which could undergo cyclization and loss of water to form the anhydride. Monomer then adds to the anhydride by a normal nucleophilic substitution reaction to form the polymer. If the reaction sequence leading to anhydride formation were reversible in favor of the acid, it could lead to low molecular weight products, as has been found, because of the reduced activity of the acid compared to the anhydride. Hydrolysis of the anhydride or deactivation of the growing polymer at the electrode surface could also lead to low molecular weight products. The thermal stability of the polymer at temperatures up to 400 °C, was comparable to that of polyamide-imide polymers. Since the electrolyte was also a solvent for the polymer, only a very thin coating was deposited on the anode.

These studies were extended to other derivatives of phthalic acid: 4-(4'-amino-benzamido), 4-(2'-aminobenzoyl), and 4-(3'-aminobenzoyl)-phthalic acid[68]. As with the 4-aminophthalic acid, only low molecular weight materials were obtained by electropolymerization of these derivatives. However, polymerization in ethanol produced a heavy coating on the anode while that in acetamide produced only a very thin coating. The amide acid polymer could be converted to the imide by heating, but the film formed was brittle.

III Electropolymerization on Graphite Fiber Electrodes[6, 7]

The interest in polymer coating of graphite fibers by electropolymerization arose out of the use of these fibers in reinforced polymer composites. In order to improve the poor bonding between the matrix polymer and unmodified graphite fiber and, thereby, the interlaminar shear strength of the composite, various methods of surface modifications, including surface oxidation of the fibers, had been developed. However, surface treatments which increase fiber-matrix adhesion generally lowered the impact resistance of the composite. In this context, it was reasoned that, by suitably forming and controlling the properties of the resin interlayer between the fiber reinforcement and matrix, it should be possible to modify crack initiation and propagation in the composite. In our investigation, electrochemical polymerization seemed to be an attractive process for forming a polymer layer of controlled thickness and properties on carbon fibers before their incorporation in a resin matrix.

a) Polymer Coating of Graphite Fibers

Using commercially available graphite fibers, it was found that graphite fibers were indeed a very good substrate for electropolymerization[6]. Polymer coatings, many visually observable, formed quickly within seconds after application of a current. Polymer presence was observed by weight increases of the fibers after polymerization, scanning electron micrographs, and, when possible, ir spectral analysis of polymer extracted from the fibers. The coatings showed considerable variation in appearance from rough to smooth.

The major step in conducting these polymerizations appears to be the selection of a solvent-electrolyte system which is capable of forming a solution with the monomer and which has sufficient current conducting properties. N,N-Dimethyl-formamide and dimethyl sulfoxide proved very useful in this respect as solvents. Both homo- and copolymerizations in aqueous and nonaqueous solvent systems were observed to form coatings. It was readily observed that monomers containing a variety of functional groups, terminal vinyl, carboxylic acid, anhydride, epoxy, and aziridinyl were observed to form coatings. It is interesting that even a liquid copolymer, VTBN (vinyl terminated butadiene-co-acrylonitrile), could be further polymerized by this technique.

Polymerizations conducted so far have produced polymer coatings with a wide range of chemical and mechanical properties which could be used to modify the carbon fiber-polymer matrix interphase of composites prepared from fibers treated by electropolymerization. A flexible polymer interphase could be introduced using systems such as methyl acrylate or VTBN, while stiffness could be achieved by using the styrene or styrene-acrylonitrile systems. Methyl methacrylate and styrene monomers would be expected to yield linear uncrosslinked polymer coatings while monomers which are multifunctional such as N,N'-methylenebisacrylamide or multifunctional aziridines would be expected to crosslink in the systems they were used in.

b) Polymer Grafting to Graphite Fibers

Chemical bonding of the polymer to the fiber was investigated by weight increase of fibers after an electropolymerization treatment and extraction in a known solvent, followed by scanning electron microscopy. Untreated carbon fibers were subjected to electropolymerization followed by continuous extraction for a period of 120 hours to ensure removal of all the unbonded polymer which would be soluble. Observed weight increases were used as preliminary evidence of the presence of a grafted polymer. Scanning electron micrographs of the extracted fibers confirmed the presence of unextractable polymer. In the case of monomers such as methyl methacrylate which would be expected to yield linear, soluble polymers, the retention of insoluble polymer on the fiber surface can be taken as strong evidence of grafting, though it cannot be useful as evidence for grafting in the case of polyfunctional monomers that would lead to network polymers. It was seen thus that diacetoneacrylamide, methyl methacrylate and styrene did form graft polymers on graphite fibers. When grafting occurred, it was observed with both high modulus and low modulus graphite fibers.

It is not unexpected that grafting should occur on graphite fiber, exposed as it is to the electropolymerization medium containing monomer during the generation of reactive species by electron transfer at the electrode surface. Organic functional groups such as —COOH and —OH, even if only in traces, are present on carbon fibers. These groups are capable of forming free radical sites, for example, by chain transfer. Initiation of free radical polymerization or termination by combination of growing polymer radicals at these sites could lead to the observed polymer grafting.

The formation of grafted polymer on the graphite fiber assumes added signif-icance in the possible effects it can have on composite properties. For example, Brie and co-workers increased the shear strength of polyester and epoxy resin composites containing graphite fibers by grafting to the fiber a carboxylated polymer compat-ible with the matrix resin or an elastomeric polymer[70, 71]. In their experiments, the fibers were treated with ozone to form surface carboxyl groups prior to the grafting reaction with monomers under gamma radiation.

c) Polymer Stereoregularity

Apart from the variations in the electron transfer properties of the electrodes, a significant factor in the role of the electrode in electrosynthesis is the adsorption of reactants and products[72]. It is conceivable that stereoregularity in the polymer could be introduced by polymerization of monomer adsorbed on graphite fibers or if the monomer were oriented on approaching the field of the electrode. Methyl meth-acrylate monomer was selected to examine the occurrence of stereoregulation because its NMR spectra are well defined and it is known that α-methyl resonances for iso-tactic (I), syndiotactic (S), and heterotactic (H) triads appear at $\tau = 8.67$, 8.90 and 8.79 respectively.

Poly(methyl methacrylate) (PMMA) extracted with chloroform from Fortafil 3T and Fortafil 5T graphite fibers which had undergone a five second polymerization was studied in a chloroform solution using tetramethylsilane (TMS) as an internal standard. A five second polymerization time was used since surface regulation can be expected to extend only to the first layers of polymer on fiber surfaces. Due to the small amounts of PMMA obtained by extraction, the concentration of the poly-mer was unknown. High modulus (5T) and lower modulus (3T) graphite fibers were used because some differences can be expected in their surface characteristics due to differences in graphitization in the two fibers.

Three solvent-electrolyte systems were studied in two of which the locus of poly-merization was at the cathode and in one, at the anode. Intensities of the α-methyl proton resonances for I, S, and H triads were measured using a 270 Hz full scale scan over the appropriate region of the spectra. 1080 Hz full scale scans were used along with infrared spectral data to confirm the presence of PMMA.

Experimentally obtained values for the I, S, and H triads in terms of percent composition are shown in Table 2. Also included in this table are values S^* and H^* for syndiotactic and heterotactic triads calculated on the basis that the probability σ of forming isotactic triads is equal to the experimental I values; it was also assumed, that the configuration of the monomer added is independent of the configuration of

Table 2. The stereochemical configuration of electropolymerized PMMA

Solvent-electrolyte electrode system	Fortafil fibers	Percent composition of triads				
		I	S	H	S^*	H^*
DMF/NaNO$_3$/Cathode, 25 °C	3T	38	10	52	15	47
	5T	38	16	46	15	47
DMSO/NaNO$_3$/Anode, 25 °C	3T	34	17	49	17	49
	5T	43	12	45	12	45
CH$_3$OH/LiOAc/Cathode, 25 °C	3T	44	19	37	11	45
	5T	40	15	45	13	47

the monomeric unit at the growing end of the polymer and that therefore, the probabilities for syndiotactic and heterotactic triads are $(1 - \sigma)^2$ and $2(\sigma - \sigma^2)$ respectively. The S^* and H^* values calculated by this simple statistics are identical with the experimental S and H values for anodic polymer and only slightly different for the cathodic polymer. Therefore, although there is a greater number of isotactic triads than syndiotactic ones in the polymer, the experimental evidence is not conclusive that graphite fiber surface exerts a stereochemical influence during electropolymerization on the graphite electrode. In contrast to these observations on PMMA formed on the electrode are the results obtained with PMMA formed in solution. The latter was obtained by precipitation with methanol from electrolytic solutions in N,N'-dimethyl formamide and dimethyl sulfoxide after a 72 hours polymerization time. The I, S, and H triad compositions of this polymer were 25, 25, and 50 percent respectively. A 25:25:50 percent composition for I, S, and H triads would be the expected composition for a completely atactic polymer. Perhaps the increased proportion of isotactic placement in the electrode polymer results from some preferential monomer adsorption on the graphite surface facilitating isotactic addition. These aspects of electropolymerization need to be investigated thoroughly. Considerably more sophisticated techniques will be needed because of the associated experimental difficulties. The isolation of the polymer formed initially on the surface has proved difficult because of the small quantities involved and also because it is partly grafted to the graphite surface. The applicability of NMR is also limited to the few cases of polymers whose spectra are unambiguously defined for the various stereochemical configurations of the repeating units.

It is relevant to draw attention here to a related observation of Bogenschütz in studies of electropolymerization of diacetoneacrylamide[32]. In this study, the electropolymerized layer of polymer on metal electrodes was examined by X-ray diffraction. As mentioned earlier, it was observed that the diffuse halo of the polymer changed to sharp lines on annealing the polymer above the melting point at 200 °C, indicating that stereoregular polymer was formed in electropolymerization. On the other hand, Bruno, et al.[73], electropolymerized acrylonitrile in dichloromethane solution on an iron electrode and, using carbon-13 NMR, could not detect any evidence for stereoregularity in the polymer.

d) Composites of Coated Fibers

The principal effect of electropolymerization of monomers on graphite fibers is
expected to be in altering the interfacial bond strength of the coated fibers when
incorporated in a composite. Initially, it was attempted to study the adhesion of
the polymer-coated graphite fibers to an epoxy matrix by the single fiber pull-out
test[7]. The method was not capable of detecting the variability in improvement of
adhesion caused by the different types of polymer coating produced on the fibers
by electropolymerization. However, when composites were prepated by incorporation of the electrolytically coated fibers in an epoxy matrix, variations in interlaminar shear (ILS) and impact strengths were observed.

Some illustrations are shown in Table 3. Shear strengths of composites prepared from fibers treated by a 2.5 second electropolymerization of selected systems
are shown in this table. It can be seen from these figures that incorporation of a
polymer interlayer on graphite fibers prior to embedding them in a polymer matrix
has significantly affected the interlaminar shear strength of the composite. Although
further detailed studies are needed to standardize the electropolymerization technique in order to obtain optimum results, a number of experimental parameters

Table 3. Effect of electroinitiated polymer coating on interlaminar shear strength of graphite
fiber composites

Electropolymerization system[a]	V_f[b] in %	G/MPa[c]	σ/MPa[d]
None (AU)[e]	47.5	69.5	0.86
	49.3	65.6	1.9
	64.5	68.1	2.2
	67.0	69.9	0.69
	72.0	71.6	1.2
DAA/H_2SO_4	54.9	71.7	0.59
	65.0	75.7	1.8
	68.0	75.2	1.7
Methyl methacrylate NaNO$_3$/DMF	47.0	61.3	3.3
	61.8	63.8	1.3
	69.8	70.0	1.1
Acrylic acid/H_2SO_4	46.8	73.6	2.1
	62.3	78.2	2.2
	64.0	79.0	5.0
	69.8	61.2	2.5
DAA/H_2SO_4 (10 s)	49.8	81.8	2.2
	64.3	78.5	4.1
	67.3	71.2	4.3

[a] 2.5 s polymerization.
[b] V_f = fiber volume fraction.
[c] G = interlaminar shear strength.
[d] Standard deviation of G.
[e] Hercules AU untreated fiber.

have been identified. The resulting composite mechanical properties were found to be a function of the monomer, solvent, polymerization time, fiber content, and post electropolymerization treatment of the coated carbon fibers.

Since significantly different shear strengths are obtained using different types of monomers in electropolymerization, it would appear that the shear strength is quite sensitive to the chemical and structural properties of the electrolytically formed polymer interphase. The results of impact strengths on notched specimens provided further evidence that the carbon fiber-polymer matrix interface could be modified by electropolymerization.

It is necessary to recognize here that in this study the composite specimens were prepared from fibers which were coated by polymers formed under one set of electropolymerizations conditions, and that these conditions were not optimized with respect to composite properties. The results are thus taken only to indicate the potential of electropolymerization for interphase modification in graphite fiber composites and the need to standardize electropolymerization conditions and monomer system to control polymer film properties and, through them, composite properties.

An important extension of the electropolymerization experiments has now been made to include polyimide precursors[74]. Polyimide intermediates have been recently synthesized with thermally cross-linkable vinyl, cyano, and acetylenic groups through which curing can be accomplished without the generation of volatiles[75-77]. Acetylene terminated polyimide oligomer HR-600 (Gulf Chemical Co.), for example, has been shown in our laboratories to electropolymerize on graphite fibers[74]. The electropolymerization of 4-aminophthalic acid[67, 68] discussed earlier is also being investigated on graphite surfaces. The electropolymerization studies of such precursors of high temperature resistant matrix polymers could provide significant advances in composites science and technology. Electropolymerization of monomers could thus be an excellent complement to electrodeposition of polymers on graphite fibers which has also been shown to be an extremely effective method of improving shear and impact strengths of graphite composites by interphase modification[78].

e) Phenylacetylene

The polymerization of cyano- and acetylene-terminated polyimide precursors suggested the need to establish the occurrence of electropolymerization through these functional groups. The electroinitiated polymerization of benzonitrile through the nitrile group to give $-C=N$ conjugated linear polymers, $+(C_6H_5)C=N+_n$, was described earlier. Similarly, phenylacetylene was chosen as a model compound to show electropolymerization occuring through acetylene groups[10] which had not been shown in any study before.

Phenylacetylene was polymerized in the cathode compartment of a two compartment electrolytic cell, separated by a fritted glass from the anode compartment. Sodium nitrate in DMF was used as supporting electrolyte. Carbon fiber electrodes were employed in these experiments.

Table 4. Cathodic polymerization of phenylacetylene

Voltage applied in V	Yield (cathode) in %	Yield (anode) in %	M_v^a(cathode)	M_v^a (anode)
6	17.0	3.0	3100	3100
12	35.0	4.0	3000	3100
18	45.0	9.0	3200	2900
24	52.0	11.0	3300	3000
36	36.0	23.0	3200	3000
48	37.0	27.0	3300	3000

[a] M_v = viscosity average molecular weight.

The polymer from phenylacetylene was formed in the cathode compartment with the development of a deep red color in $NaNO_3$/DMF solutions, but no reaction was observed when the monomer was present in the anode compartment. The migration of the color to the anode compartment from the cathodic solution was also noted. These observations, coupled with the high sensitivity to moisture which completely inhibited the polymerization, suggest an anionic mechanism of polymerization.

The polymer was deposited on the cathode and was also present in solution. Even though the reaction was conducted for a long time to collect enough polymer for characterization, polymer formation was visually observable in only 5 minutes. From viscosity measurements of the red-orange polymer isolated from the reaction mixture, the avarage molecular weight was found to be 3000. Fractional precipitation by methanol and methanol-water separated the polymer into fractions of molecular weight from 4000 to 2700 which had identical NMR and IR spectra though different melt-softening temperatures (155–167 °C, 110–112 °C, 55–76 °C). X-ray diffraction did not show any crystallinity. The results are summarized in Table 4.

Carbon hydrogen analyses agreed well with calculated values for poly(phenylacetylene) (calculated for C_8H_6: C, 94.08; H, 5.92. Found: C, 92.14–93.05; H, 5.75–5.84). The ir spectrum showed absorptions due to stretching vibrations of poly-conjugated double bonds at 1590 cm^{-1} and those due to C–H out-of-plane deformation of monosubstituted benzene at 750 cm^{-1} and 690 cm^{-1}. Weak absorptions are also seen at 910 cm^{-1} and 840 cm^{-1}, the larger of these being at 910 cm^{-1}. This region is believed to be involved in the unsaturation in poly(phenylacetylene)s, with characteristic absorptions arising at 910 cm^{-1}, 870 cm^{-1} and 840 cm^{-1} [79–81]. Of these three absorption bands, the one at 870 cm^{-1} has been taken as characteristic of a cis-structure[79]; it is seen from Fig. 4 that this band is not present conspicuously in the IR spectrum of the polymer prepared in this study. Based on the observations of Kern[80] who has prepared and compared poly(phenylacetylene)s from several types of chemical initiators, it can be surmised that the electroinitiated polymer is trans-rich because of cis-trans isomerization promoted by the polar solvent medium, DMF.

The ^{1}H NMR spectrum of the polymer was taken in CDCl$_3$ using TMS as an internal standard. The spectrum consists of a single broad multiplet centered about 7.0 ppm. This is consistent with published ^{1}H NMR spectra of poly(phenylacetyl-

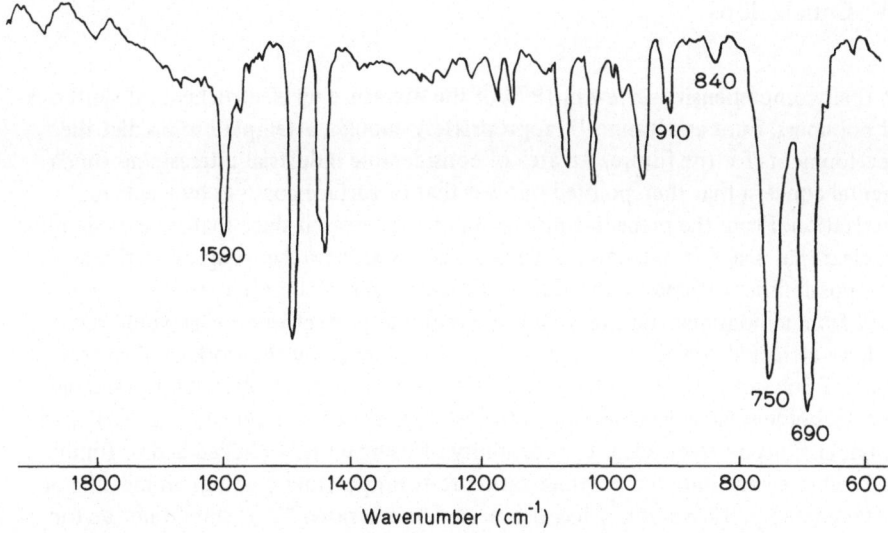

Fig. 4. IR spectrum of electroinitiated polymer of phenylacetylene

ene)[81]. Also, in compounds like 1,4-diphenyl-1,3-butadiene the $-C=C-H$ absorption in the NMR occurs at about 6.6 to 6.8 ppm[82]. An extended chain, and more likely a 1,3 placement of phenyl groups as in the more probable head-to-tail structure could cause a broadening of the phenyl and $-C=C-H$ proton absorptions and perhaps also cause chemical shift of the $-C=C-H$ protons to higher ppm values. This could lead to their being included in the benzene proton absorption resulting in a single broad multiplet about 7.0 ppm.

The available evidence then indicates the polymer to be a linear polymer with a polyene structure $(C=CH)_n$. The polymer as suggested earlier, is probably formed by an anionic mechanism at the cathode through the opening of $C\equiv C$ bonds. The presence of polymer at the anode, as seen in Table 4, increases with increasing voltage, and probably results from electrophoretic migration of living polymer anions since no monomer was present in the anode compartment. It is significant that the molecular weights of the polymers from both compartments are the same.

It would be most interesting to investigate the mechanism of electropolymerization of acetylenic monomers which has found independent confirmation in the work of Simionescu and co-workers[83] which appeared at the same time as our report[10]. Using perchlorates as electrolytes in DMF and DMSO, these authors have electropolymerized phenylacetylene and diphenyldiacetylene and obtained polymers of much lower molecular weights, 500–1000. The ir spectrum showed the formation of conjugated bonds through the polymerization of triple bonds. Interestingly, the early evidence seems to indicate that only one of the two triple bonds of diphenyldiacetylene is polymerized to give an oligomer of the structure

$$(C_6H_5)C=C]_n$$
$$\quad\quad |$$
$$\quad\quad C\equiv C(C_6H_5)$$

IV Conclusions

In their comprehensive review in 1975 of the literature on electrochemical synthesis of polymers, Funt and Tanner[1], appropriately enough, attempted to predict the developments for the future. An area of considerable industrial interest and fundamental concern that they pointed out was that of surface coatings by electropolymerization. From the preceding discussion of the research since that time, it should be clear that a significant start has been made towards further progress in this area. The possibilities offered by the ability, via electrochemical polymerization, to proceed from a monomer solution to an adherent and protective coating would seem to have attracted considerable attention — as evidenced by the work on the formation of poly(arylene oxide) and polyimide coatings by electroinitiation. Valuable new techniques have been developed to investigate the formation and properties of polymer films on electrodes. The feasibility of using either a packed bed or fluidized bed cell in electroinitiated polymerization to form polymer coatings on metallic or metal-coated particle surfaces has also been demonstrated[84]. It should not be too long before the technological promise of these investigations is realized.

Electropolymerization on graphite fibers signifies a new development. It can be expected that the understanding of the chemical and mechanical requirements of the interphase, and of the mechanisms of reinforcement in graphite composites will be enhanced by investigations of interphase modification by the valuable new technique available now based on electrodic processes. The control of stereoregularity in electrochemical polymerization invites further careful study. Finally, exciting new avenues have been opened up in the study of the mechanisms of electropolymerizations that have now been shown to occur through acetylenic and nitrile groups.

Acknowledgement. It is a pleasure to acknowledge the contributions of my students and associates whose work from the cited references has been so extensively reviewed in this paper, particularly Drs. Brij K. Garg and James J. Jakubowski.

V References

1. Funt, B. L., Tanner, J.: Tech. Chem. (N.Y.) *5* (2), 599 (1975)
2. Shapoval, G. S., Gorodyskii, A. V.: Usp. Khim. *42* (5), 854 (1973)
3. Breitenbach, J. W., Olaj, O. F., Sommer, F.: Fortschr. Hochpolym.-Forsch. *9*, 47 (1972)
4. Parravano, G. J., in: Organic Electrochemistry, p. 947. Baizer, M. N. (ed). New York: Marcel Dekker, 1973
5. Garg, B. K., Raff, R. A. V., Subramanian, R. V.: J. Appl. Polym. Sci. *22*, 65 (1978)
6. Subramanian, R. V., Jakubowski, James J.: Polym. Eng. Sci. *18*, 590 (1978)
7. Subramanian, R. V., Jakubowski, James J., Williams, F. D.: J. Adhesion *9*, 185 (1978)
8. Subramanian, R. V., Garg, B. K., Doun, Fon-san, Mahalingham, R.: Polym. Prepr., Am. Chem. Soc. Div. Polym. Chem. *18* (2), 420 (1977)
9. Teng, F. S., Mahalingham, R., Subramanian, R. V., Raff, R. A. V.: J. Electrochem. Soc. *124*, 995 (1977)

10. Subramanian, R. V., Jakubowski, J., Garg, B. K.: U.S. NTIS, A. D. Rep. No. AD-A047492 (1977); Chem. Abstr. *88*, 170537q (1978)
11. Ross, S. D., Kelly, D. J.: J. Appl. Polym. Sci. *11*, 1209 (1967)
12. Asahara, T., Seno, M., Tsuchiya, M.: Kinzoku Hyomen Gijutsu *19*, (12), 511 (1968)
13. Idem: Kinzoku Hyomen Gijutsu *20* (8), 411, 414 (1969)
14. Idem: Kinzoku Hyomen Gijutsu *20* (1), 2, 28 (1969)
15. Idem: Kinzoku Hyomen Gijutsu *20* (2), 64 (1969)
16. Idem: Kinzoku Hyomen Gijutsu *20* (3), 99 (1969)
17. Idem: Kinzoku Hyomen Gijutsu *20* (11), 576 (1969)
18. Asahara, T., Seno, M., Tobayama, M.: Proc. Congr. Int. Union Electrodeposition Surf. Finish. *8*, 254 (1973)
19. Bezuglyi, V. D., Korshikov, L. A., Kravatsova, L. I., Bondarenko, I. B., Fioshin, M. Ya.: Elektrokhimiya *8*, (11), 1658 (1972)
20. Korshikov, L. A., Kravatsova, L. I., Bezuglyi, V. D.: Elektrokhimiya *10* (1), 106 (1974)
21. Korshikov, L. A., Karpinets, A. P., Bezuglyi, V. D.: Elektrokhimiya *10* (6), 990 (1974)
22. Bezuglyi, V. D., Karpinets, A. P., Korshikov, L. A.: Novosti Elektrokhim. Org. Soedin., Tezisy Dokl. Vses. Soveshsh, Elektrokhim, Org. Soedin *8*, 62 (1973)
23. Bezuglyi, V. D., Karpinets, A. P., Korshikov, L. A.: Vysokomol. Soedin., Ser. B: *16* (8), 601 (1974)
24. Mengoli, G., Tidswell, B. M.: Polymer *16*, 881 (1975)
25. Sobieski, J. F., Zerner, M. C.: U.S. Pat. 3,464,960 (1969); Chem. Abstr. *71*, 103294u
26. Rovinelli, R. J. (ed.): Applications of photopolymers, p. 102. Washington, D. C.: Society of Photographic Scientists and Engineers, Inc., 1970
27. Collins, G. L., Thomas, N. W.: J. Polym. Sci. Polym. Chem. Ed. *15*, 1819 (1977)
28. W. R. Grace Co.: Br. Pat. 1,134,387 (1968); Chem. Abstr. *70*, 30197g
29. W. R. Grace Co.: Fr. Pat. 1,586,798. Chem. Abstr. Patent Concordance *73*, 43pc
30. W. R. Grace Co.: Br. Pat. 1,179,543 (1970); Chem. Abstr. *72*, 101933y
31. Bogenschütz, A. F., Jostan, J. L., Krusemark, W.: Galvanotechnik *60* (10), 750 (1969)
32. Idem: Kunststoffe *60* (2), 127 (1970)
33. Krusemark, W. R., Jostan, J. L., Bogenschütz, A. F.: Fachberichte für Oberflächentechnik (Jan. 1970), p. 18
34. Jostan, J. L., Krusemark, W., Bogenschütz, A. F.: Oberfläche-Surface *10* (10), 677 (1969)
35. Panaiotov, I. M., Obreshkov, A. T.: Vysokomol. Soedin *7* (2), 366 (1965)
36. Trifonov, A. Z., Schopov, I. D.: J. Electroanal. Chem. *35*, 415 (1972)
37. Schurz, J., Bayzer, H., Stübchen, H.: Makromol. Chem. *23*, 152 (1957)
38. Dyer, J. A.: Applications of absorption spectroscopy of organic compounds, Chapt. 3. Englewood Cliffs: Prentice-Hall, 1965
39. Fester, W., in: Polymer Handbook, 2nd ed., p. V-37. Brandrup, J., Immergut, E. H. (eds.). New York: Wiley-Interscience, 1975
40. Kabanov, V. A., Zubov, V. P., Kovaleva, V. P., Kargin, V. A.: J. Polym. Sci., Part C *4*, 1009 (1963)
41. Kargin, V. A., Kabanov, V. A., Zubov, V. P., Zezin, A. B.: Dokl. Akad. Nauk SSSR *139* (3), 605 (1961)
42. Liepins, R., Campbell, D., Walker, C.: J. Polym. Sci., Part A-1 *6*, 3059 (1968)
43. Liepins, R.: Makromol. Chem. *118*, 36 (1968)
44. Wildi, B. S., Katon, J. E.: J. Polym. Sci., Part A *2*, 4709 (1964)
45. Oikawa, E., Kambara, S.: Bull. Chem. Soc. (Japan) *37*, 1849 (1964)
46. Norrell, C. J., Pohl, H. A., Thomas, M., Berlin, K. D.: J. Polym. Sci., Polym. Phys. Ed. *12*, 913 (1974)
47. Hsu, L. C.: ACS Symp. Ser. *4*, 145 (1974)
48. Weiranch, K. K., Gemeinhardt, P. G., Baron, A. L.: 34th Ann. Meeting, Soc. Plast. Eng. *32*, 317 (1976)
49. Griffith, J. R., O'Rear, J. G., Walton, T. R.: Adv. Chem. Ser. *142*, 458 (1975)
50. W. R. Grace Co.: Br. Pat. 1,156,309 (1969); Chem. Abstr. *71*, 72049m

51. Dijkstra, R., De Jonge, J.: Sci. Technol. Surf. Coat., NATO Adv. Study Inst., 1972 (Publ. 1974), 85
52. Iwakura, C., Tsunaga, M., Tamura, H.: Electrochim. Acta *17*, 1391 (1972)
53. Bruno, F., Pham, M. C., Dubois, J. E.: Electrochim. Acta *22*, 451 (1977)
54. Epimakhov, V. N., Mishenov, Yu M., Yudkin, B. I.: Plast. Massy (9), 74 (1975)
55. Schulz, R. C., Strobel, W.: Monatsh. Chem. *99*, 1742 (1968)
56. Strobel, W., Schulz, R. C.: Makromol. Chem. *133*, 303 (1970)
57. Desbene-Monvernay, A., Dubois, J. E., Lacaze, P. C.: J. Electroanal. Chem. Interfacial Electrochem. *89*, 149 (1978)
58. Funt, B. L., Rybicky, J.: J. Polym. Sci., Polym. Chem. Ed. *9*, 1441 (1971)
59. Funt, B. L., McGregor, I., Tanner, J.: J. Polym. Sci., Polym. Lett. Ed. *8*, 695 and 699 (1970)
60. Phillips, D. C., Davies, D. H., Smith, J. D.: Makromol. Chem. *154*, 32 (1972)
61. Yamazaki, N., Shinohara, H., Nakahama, S.: J. Macromol. Sci.-Chem. *A 9 (4)*, 539 (1975)
62. Kabanov, V. A., Topchiev, D. A., Karaputadze, T. M.: J. Polym. Sci., Polym. Symp. *42*, 173 (1973)
63. Ponratnam, S., Kapur, S. L.: J. Polym. Sci., Polym. Chem. Ed. *14*, 1987 (1976)
64. Ponratnam, S., Rao, S. P., Joshi, S. G., Kapur, S. L.: J. Macromol. Sci.-Chem. *A 10 (6)*, 1055 (1976)
65. Cabaness, W. R., Lin, T. Y., Parkanyi, C.: J. Polym. Sci., Polym. Chem. Ed. *9*, 2155 (1971)
66. Phillips, D. C.: J. Electrochem. Soc. *119*, 1645 (1972)
67. Phillips, D. C., Spewock, S., Alvino, W. M.: J. Polym. Sci., Polym. Chem. Ed. *14*, 1137 (1976)
68. Phillips, D. C., Alvino, W. M.: J. Polym. Sci., Polym. Chem. Ed. *14*, 1151 (1976)
69. Laube, B. L., Higgins, J.: J. Polymer Sci., Part A-1, *10*, 2389 (1972)
70. Brie, M., Cazard, J., Lang, F. M., Riess, G.: Bull. Inform. Sci. Tech. Commis. Energ. At. *155*, 31 (1971)
71. Brie, M., Legressus, C.: Fiber Sci. Technol. *6*, 47 (1973)
72. Funt, B. L.: Macromol. Rev. *13*, 37 (1967)
73. Bruno, F., Pham, M. C., Dubois, J. E.: J. Chim. Phys., Phys. Chim.-Biol. *72* (4), 490 (1973)
74. Jakubowski, J., Subramanian, R. V.: unpublished results
75. Wentworth, S. E., Macaione, D. P.: J. Polym. Sci., Polym. Chem. Ed. *14*, 1301 (1976)
76. Bilow, N., Landis, A. L.: National SAMPE Tech. Conf. Series *8*, 94 (1976)
77. Bilow, N., Landis, A. L., Miller, L. J., Lawrence, R. E., Aponyi, T. J.: Polym. Prepr., Am. Chem. Soc., Div. Polym. Chem. *15* (2), 542 (1974)
78. Subramanian, R. V., Patel, A. K., Sundaram, V.: Proc. SPI Reinforced Plast/Comp. Inst., 20F (1978)
79. Masuda, T., Sasaki, N., Higashimura, T.: Macromolecules *8*, 717 (1975)
80. Kern, R. J.: J. Polym. Sci., Part A-1, *7*, 621 (1969)
81. Sasaki, N., Masuda, T., Higashimura, T.: Macromolecules *9*, 664 (1976)
82. Aldrich: NMR Library, Spectrum No. 4, 20C
83. Farafonov, V., Grovu, M., Simionescu, C.: J. Polym. Sci., Polym. Chem. Ed. *15*, 2041 (1977)
84. Mahalingham, R., Teng, F. S., Subramanian, R. V.: J. Appl. Polym. Sci. *22*, 3587 (1978)

Received March 13, 1979

Molecular Aspects of Multiple Dielectric Relaxation Processes in Solid Polymers

Graham Williams

Edward Davies Chemical Laboratories, University College of Wales, Aberystwyth, Dyfed, SY23 1NE, U.K.

The article outlines our current understanding of the multiple relaxations observed in crystalline and amorphous solid polymers, as studied using dielectric techniques. An attempt is made to interpret the relaxations of amorphous polymers in a unified way, independent of the details of chemical structure, by use of the time-correlation function approach to partial and total relaxations. In addition, the recent studies of polymers of medium and high degrees of crystallinity are reviewed.

Table of Contents

I Introduction

Following the classic studies of Fuoss, Kirkwood and their co-workers over forty years ago, there has been a steady output of publications concerned with the dielectric relaxation of amorphous and crystalline solid polymers. Such publications have
(i) documented, over a wide range of frequency and temperature (and in some instances pressure), the behaviour of new polymer systems,
(ii) correlated dielectric, mechanical and related relaxation behaviour of well known and new polymer systems,
(iii) obtained insights into the structural and molecular factors which are responsible for the observed multiple relaxations,
(iv) obtained information on electrical relaxation in order to aid the interpretation of related phenomena, e.g. piezo-electrical and pyro-electrical behaviour and electrostriction, for speciality polymers [e.g., poly(vinylidene fluoride)].
In view of the breadth of this work, its interdisciplinary nature and its importance for polymeric materials, there will be a continuing and active interest in the dielectric relaxation of solid polymers. Whilst the phenomenological aspects of molecular relaxation in general, and dielectric relaxation in particular, are well-established (McCrum et al., 1967) there is a need to place on a firm basis our understanding of the structural and molecular factors which are responsible for the observed behaviour [i.e. (iii) above]. Substantial accounts of the dielectric behaviour of solid polymers have been given by McCrum et al. (1967), Ishida (1969), Hedvig (1977), Baird (1973), Wada (1977) and Karasz (1972). Further reviews are given by Williams and Watts (1971 a), Van Turnhout (1975, 1978) and by Williams and Crossley (1978). Heijboer (1972) has given a valuable account of relaxations for glassy polymers containing flexible groups.
The present account is not intended to be comprehensive. Data which are regarded as representative of the dielectric behaviour of amorphous and partially crystalline polymers will be discussed and the structural and molecular factors involved will be examined in some detail. A brief account will also be given of the relevance of dielectric relaxation behaviour to topics of current interest such as piezo-electrical behaviour of polymers. This account attempts to build on the many earlier accounts (McCrum et al., 1967; Ishida, 1969; Hedvig, 1977; Baird, 1973; Wada, 1977; Karasz, 1972; Williams and Watts, 1971 a; Van Turnhout, 1975, 1978; Williams and Crossley, 1978; Heijboer, 1972) and is a personal assessment by the author of our current understanding of the factors which lead to multiple dielectric relaxation processes in solid polymers.

II Amorphous Polymers

A. Introduction

Before we consider amorphous polymers in detail, we note that most dielectric studies involve measurement of the dielectric permittivity $\epsilon^* = \epsilon' - i\,\epsilon''$, where ϵ'

and ϵ'' are the real permittivity and loss-factor respectively, as a function of frequency f at given temperatures and that multiple relaxations may be observed (McCrum et al., 1967). If the relaxations are well separated then three items of information may be obtained for each process at each temperature, viz. (a) the dispersion magnitude $\Delta \epsilon$, (b) the frequency f_m of maximum loss-factor, and (c) the form of the ϵ'' vs. $\log f$ plot. Measurements at different temperatures yield the apparent activation energy Q_{app} (T) for each process. Thus interpretations of data in molecular terms are based upon $\Delta \epsilon$ (T), $\log f_m$ (T), the "line-shape" and Q_{app} (T) for each process.

B. The α-, β- and $(\alpha\beta)$-Relaxations

It is well established that all solid amorphous polymers exhibit a principal (or α) and a secondary (or β) process (McCrum et al., 1967; Ishida, 1969; Hedvig, 1977; Baird, 1973; Wada, 1977). Many polymers exhibit further relaxations: e.g., poly-(methyl methacrylate) exhibits five processes (see McCrum et al., 1967, p. 250). Not all relaxation processes are observed in a dielectrics experiment. A process is dielec-trically active only if it involves the reorientation of the dipole-moment vector[1].

For the majority of amorphous polymers $\Delta \epsilon_\alpha \gg \Delta \epsilon_\beta$ e.g., for poly(ethylene terephthalate), which contains dipoles rigidly attached to the main-chain, and poly-(vinyl acetate), which has dipole moment components rigidly attached to the main-chain and attached to the flexible side-group (see McCrum et al., 1967). For certain amorphous polymers, notably the conventional poly(n-alkyl methacrylates) $\Delta \epsilon_\beta > \Delta \epsilon_\alpha$ (McCrum et al., 1967; Williams and Edwards, 1966; Williams, 1966 b, c; Sasabe and Saito, 1968; Sasabe, 1971). In the frequency domain $\log (f_m)_\alpha < \log (f_m)_\beta$ and there is a tendency for the two processes to come together with increasing tem-perature. For most experiments the dielectric α-process has been studied above the apparent glass-transition temperature (T_g) in the frequency range 10^{-2} to 10^6 Hz. Similarly, most studies of the β-process have been made for $T < T_g$ and in the same frequency range as that of the α-process. We now summarize, with examples, the appearance and behaviour of α-, β- and $(\alpha\beta)$-processes and note that these have been briefly discussed in earlier publications (see Williams, cited in Karasz, 1972; Williams and Watts, 1971 a).

With the exception of poly(chlorotrifluoroethylene) (Scott and co-workers, 1962) polymers containing their dipoles rigidly attached to the main-chain have $\Delta \epsilon_\alpha \gg \Delta \epsilon_\beta$. Certain polymers containing flexible dipolar side-groups also have $\Delta \epsilon_\alpha > \Delta \epsilon_\beta$, e.g. poly(vinyl acetate), poly(methyl acrylate) and poly(ethyl acrylate) (see McCrum et al., 1967). Also the higher poly(n-alkyl methacrylates) have $\Delta \epsilon_\alpha > \Delta \epsilon_\beta$ e.g. poly(n-nonyl methacrylate) and poly(n-lauryl methacrylate) (Wil-liams and Watts, 1971 b). However, certain polymers containing flexible dipolar side groups have $\Delta \epsilon_\beta > \Delta \epsilon_\alpha$ at certain temperature-pressure conditions: e.g. "atactic"

---

1 *Note:* Space-charges may also give active dielectric processes. These are commonly observed as a limiting low-frequency process (McCrum, Read and Williams, 1967; Baird, 1973) or as a high-temperature peak in a thermally stimulated depolarization experiment (Van Turnhout, 1975, 1978). In this article we shall be mainly concerned with the intrinsic molecular behav-iour arising from dipole motion.

poly(methyl methacrylate) and poly(ethyl methacrylate) (McCrum et al., 1967; Williams and Edwards, 1966; Williams, 1966 b; Sasabe and Saito, 1968). $\Delta \epsilon_{total}$ = = $\Delta \epsilon_\alpha + \Delta \epsilon_\beta$ at a given temperature. $\Delta \epsilon_{total} \propto C_r \langle \mu^2 \rangle$ where C_r is the concentration of dipolar groups and $\langle \mu^2 \rangle$ is the apparent mean-square dipole moment of a group. Thus the strength of relaxation is partitioned between the α- and β-processes. For $\Delta \epsilon_\alpha \gg \Delta \epsilon_\beta$ the β-process relaxes only a small part of $\langle \mu^2 \rangle$, the remainder being relaxed by the α-process (if it occurs at a finite rate at the temperature of measurement). The converse is true for $\Delta \epsilon_\alpha \ll \Delta \epsilon_\beta$. Most workers would agree that in general terms the α-process corresponds to the gross-microbrownian motions of chains (and is thus the "dynamic glass-transition" of the polymer) while the β-process corresponds to limited motions of chains. We know that dipoles are moving, we know how fast they are moving on average (via the average relaxation times $\langle \tau_\alpha \rangle, \langle \tau_\beta \rangle$) and we know the general origins of α- and β-processes. However, many models have been proposed for the way the chains move to give the α- and β-processes, and there has been much uncertainty regarding the mechanisms for many years. This will be discussed in Sections IIC and IID and it will be shown that it is not possible to establish a mechanism for motion using the results of a single experimental technique. Additional evidence is required in order to establish the mechanism and this can only be obtained from the use of complementary experimental methods.

The dielectric α process is well defined in the frequency domain for a large number of polymers (McCrum et al., 1967). The complex permittivity $\epsilon^*(\omega)$ in the frequency region of a well defined α-process may be related to the relaxation function $\Gamma_\alpha(t)$, say, by a Fourier transformation (Williams, 1972 a; Williams, 1978).

$$\frac{\epsilon^*(\omega) - \epsilon_{\infty \alpha}}{\epsilon_{0\alpha} - \epsilon_{\infty \alpha}} = 1 - i \omega \int_0^\infty \Gamma_\alpha(t) \exp(-i\omega t) dt \qquad (1)$$

$\epsilon_{0\alpha}$ and $\epsilon_{\infty \alpha}$ are the limiting low and high frequency permittivities respectively, ω is the angular frequency, $\omega = 2 \pi f$. If the relaxation function is given by the empirical relation of Williams and Watts 1970; Williams et al., 1971)

$$\Gamma_\alpha(t) = \exp - (t/\tau_0)^{\bar{\beta}} ; \ 0 < \bar{\beta} \leqslant 1 \qquad (2)$$

then Eqs. (1) and (2) give curves of $\epsilon''(\omega)$ against log ω which are nonsymmetrical about the maximum loss condition. It is found that the relaxations for polymers of quite different chemical structure, containing dipoles both rigidly attached and flexibly attached to the main chain, may be fitted using this empirical relaxation function with $0.4 < \bar{\beta} < 0.6$, e.g. styrene-acrylonitrile copolymers, styrene-o-chlorostyrene copolymers, poly(vinyl acetate), poly(methyl acrylate), poly(ethyl acrylate), poly(vinyl octanoate), poly(ethylene terephthalate), poly(propylene oxide) and many others (Williams and Watts, 1971 a; Williams et al., 1971; Williams et al., 1972a). This overall similarity for the shape of the α-process for such widely differing polymers implies a common specific mechanism for motion.

The dielectric β-process gives very broad loss curves in the frequency domain for all amorphous polymers, the half-width of the curves being 4–6 decades of frequency. Whilst various empirical representations (see e.g., McCrum et al. 1967; Williams and Watts, 1970; Williams et al., 1971) may be used to fit such data, including the use of distributions of relaxation times (McCrum et al., 1967) it seems difficult to extract information on basic mechanisms for motion when processes are as broad as this. Clearly the observed process for a given polymer may be represented by a variety of mechanisms each with its own adjustable parameters.

With regard to the temperature dependencies of $\langle \tau_\alpha \rangle$ and $\langle \tau_\beta \rangle$, where $\langle \tau_i \rangle = \log (f_m)_i$, it is well-known that plots of $\log \langle \tau_\alpha \rangle$ against T^{-1} are strongly curved in approximate accord with the Williams-Landel-Ferry relation while the corresponding plot for $\log \langle \tau_\beta \rangle$ is linear in accord with the Arrhenius relation (Ferry, 1961; McCrum et al., 1967). In recent years several studies have been made of the effect of a hydrostatic pressure on both α- and β-processes, and some of this work has been reviewed (see Williams, in Karasz, 1972, p. 17; Williams and Watts, 1971 a; Sasabe, 1971). Poly(vinyl acetate) (O' Reilly, 1962), polyvinyl chloride (Koppelman and Gielessen, 1961; Williams and Watts, 1971c; Saito and co-workers, 1968 and Sasabe 1971), poly(ethylene terephthalate) (Williams, 1966 a; Saito and co-workers, 1968), poly(methyl acrylate) (Williams, 1964 a, b) poly(ethyl acrylate) and poly(ethylmethacrylate) (Williams, 1966 b), poly(methyl, ethyl, n-butyl, n-octyl, and n-lauryl methacrylates) (Sasabe and Saito, 1968; Sasabe, 1971), poly(n-nonyl methacrylate) and poly(n-lauryl methacrylate) (Williams and Watts, 1971 b), acrylonitrile-butadiene copolymers (Williams et al., 1972b) and poly(propylene oxide) (Williams, 1965) have been studied. The general pattern of behaviour is as follows. The α-process moves rapidly to lower frequencies with increasing pressure, with $[\partial \log f_{m\alpha}]/\partial P)_T$ in the range 1×10^{-3} to 4×10^{-3} atm^{-1} [2], and without an appreciable change in $\Delta \epsilon_\alpha$ or $\bar{\beta}$ (i.e. for those polymers for which $\Delta \epsilon_\alpha \gg \Delta \epsilon_\beta$). Studies of the β-process at $T < T_g$ indicate that in this range $\Delta \epsilon_\beta$, $\bar{\beta}$ and $\log (f_m)_\beta$ are rather insensitive to pressure. $(\partial \log (f_m)_\beta / \partial P)_T$ lying in the range 0.1×10^{-3} to 1×10^{-3} atm^{-1}. The behaviour of flexible side-group polymers [e.g. the poly(alkyl methacrylates)] is similar to that for polymers such as poly(vinyl chloride) and poly(ethylene terephthalate) when $T < T_g$. However, in the glass-transition range and for $T > T_g$ the β-process in the flexible side-group polymers exhibits remarkable and complicated behaviour with temperature and pressure (Williams, 1966 b; Sasabe and Saito, 1968; Williams and Watts, 1971b; Sasabe, 1971). This behaviour is closely linked to that of the $(\alpha\beta)$- and α-processes in the vicinity of T_g, so we include a discussion of it in our account of the nature and occurrence of the $(\alpha\beta)$-process.

It is clear from the studies of poly(ethyl methacrylate) (Williams, 1966b; Sasabe and Saito, 1968; Sasabe, 1971) and poly(n-butyl methacrylate) (Williams and Edwards, 1966; Sasabe and Saito, 1968; Sasabe, 1971) that the α- and β-processes coalesce above T_g to form a single $(\alpha\beta)$-process. Increase in pressure at a given temperature leads to a decomposition of the $(\alpha\beta)$-process into α- and β-processes [3]. Several important features emerge from these studies and suggest mechanisms for α-, β- and $(\alpha\beta)$-processes for the alkyl methacrylate polymers and amorphous polymers generally.

---

2 In SI-units: 1 atm = 101 325 N $\cdot$ m^{-2}.

3 *Note:* For poly(n-nonyl methacrylate) the principal relaxation is an $(\alpha\beta)$-process and this was demonstrated to be so via high pressure studies (Williams and Watts, 1971 b).

(i)   The coalescence of α- and β-processes to form the $(\alpha\beta)$-process implies that all
       three processes are inter-related.

(ii)   The coexistence of α- and β-processes (in plots of ϵ'' against $\log f$) in a tempera-
       ture range above T_g implies that the β-process is not a special feature of the
       glassy state (where it is normally studied).

(iii) The remarkable decrease in $\Delta\epsilon_\beta$ with increasing pressure above T_g for poly-
       (ethyl methacrylate) and poly-(n-butyl methacrylate) is accompanied by a cor-
       responding increase in $\Delta\epsilon_\alpha$, (Williams, 1966b; Sasabe and Saito, 1968; Sasabe,
       1971). This implies that α- and β-processes are interrelated via a conservation
       relation $\Delta\epsilon = \Delta\epsilon_\alpha + \Delta\epsilon_\beta$ with respect to changes in pressure at a given tempera-
       ture. The decrease in $\Delta\epsilon_\beta$ was interpreted (Williams, 1966b) to mean that for
       these flexible side-chain polymers the motions of the side groups, which are a major
       contributor to the β-process, were readily "blocked" by a relatively small decrease
       in local volume (see Heijboer, 1965).

(iv)   The $(\alpha\beta)$-process is to be regarded as a continuation of the α-process to higher
       temperatures, the only difference between α- and $(\alpha\beta)$-processes being that the
       the α-process relaxes a part whereas the $(\alpha\beta)$-process relaxes the whole of $\langle\mu^2\rangle$
       (Williams, 1966b; Sasabe and Saito, 1968).

Thus the α-, β- and $(\alpha\beta)$-processes have been demonstrated in occur for certain
poly(alkyl methacrylates) and their inter-relations and general mechanisms appear to
have been established. No equivalent systematic study has been made for the multiple
relaxations of other amorphous polymers. It is very important to see if the same pattern of multiple relaxations of other amorphous polymers — in particular for polymers containing dipoles rigidly attached to the main chain, e.g. poly(vinylchloride)
and poly(ethylene terephthalate). Inspection of the plots of $\log f_m$ against T^{-1}
for the multiple relaxations of many polymers (see McCrum et al., 1967; McCall,
1969) shows that for the majority of polymers the α-process was studied above
T_g and the β-process was studied below T_g. Since the locus of the α-process is curved
(WLF behaviour) an extrapolation of α- and β-processes to higher temperatures is to
be regarded as tentative. However these data suggest that α- and β-processes would
coexist for $T > T_g$ for poly(ethylene terephthalate), poly(vinyl chloride), poly(methyl
α-chloroacrylate), poly(methyl acrylate), poly(vinyl acetate) and polycarbonate. For
poly(ethylene terephthalate) and poly(vinyl chloride) we have the following additional information. Ishida (see McCrum et al., 1967, p. 508) and Saito (1964) studied
amorphous poly(ethylene terephthalate) in the range $86-102$ °C ($T_g = 67$ °C for
the amorphous polymer). The α-process was observed having $\log f_m$ in the range
10^{-1} to 10^5 Hz, dependent upon temperature. The rising loss at higher frequencies
is clear evidence that the β-process is at higher frequencies and therefore coexists
with the α-process in this temperature range. Ishida (see McCrum et al., p. 423)
studied poly(vinyl chloride) in the range -61 °C to 120 °C. The α- and β-processes
were observed in the frequency range 10^2 to 10^6 Hz. The data indicate that as the
temperature is lowered towards T_g an $(\alpha\beta)$-process resolves into a (large) α-process
and a (small) β-process. This is substantiated by the data of Sasabe (1971) and Saito
and co-workers (1968) for poly(vinyl chloride) above T_g. Increase of pressure up to
1440 atm at 103 °C clearly decomposes a single $(\alpha\beta)$-process into α- and β-processes.

The resolved β-process has the same characteristics as those for the β-process observed below T_g at atmospheric pressure (see Saito and co-workers, 1968, Fig. 4 and McCrum et al., 1967, p. 435). These data for poly(ethylene terephthalate) and poly(vinyl chloride) also suggest that α- and β-processes coalesce at higher temperatures.

Thus we suggest that there is good evidence for the existence of α-, β- and ($\alpha\beta$)-relaxations for polymers other than flexible side-chain polymers. We further suggest that this is the pattern of behaviour for all amorphous solid polymers. We do not regard this as surprising since we expect it from our general approach to dipole relaxation (see Section IIDa below) and from the understanding of the general mechanisms of the equivalent processes that have been observed for supercooled molecular liquids and glasses (Johari and Smyth, 1969, 1972; Johari and Goldstein 1970a, b; Johari, 1973; Johari, 1976; Johari, 1972; Williams and Hains, 1971, 1972; Williams, 1975). This will be further discussed below.

C. Molecular Theory of Dipole Relaxation

The theory of the static and dynamic permittivity of dipolar media has been developed over the past fifty years and many texts are available. For summaries and accounts of the current state of the theoretical work the reader is referred to the texts of Hill and co-workers (1969), McCrum et al. (1967) and Böttcher (1973, 1978). The theory of the static permittivity is so well-documented that we shall simply give here the basic relations and the definitions of terms.

The static permittivity ϵ_0 for a dipolar polarizable medium is related to the dipole moment of a macroscopic sphere of volume V according to the relation (Fröhlich, 1958; Cook et al., 1970).

$$(\epsilon_0 - \epsilon_\infty) = \frac{4\pi}{3kT} \frac{3\epsilon_0 (2\epsilon_0 + \epsilon_\infty)}{(2\epsilon_0 + 1)^2} \frac{\langle M(0) \cdot M(0) \rangle}{V} \tag{3}$$

ϵ_∞ is the limiting high frequency permittivity, $M(0)$ is the instantaneous dipole moment of the macroscopic sphere, $\langle M(0) \cdot M(0) \rangle$ is the mean square dipole moment, where the average is taken over all configurations of the ensemble contained in the sphere and is deduced in the absence of an applied field. If the volume V contains N equivalent polymer chains and the instantaneous dipole moment of a chain i is given by $P_i(0)$, where $P_i(0)$ is the vector sum of the dipole moments $\mu_{ji}(0)$ of the chain i, then in the absence of orientational correlations *between* the dipoles of different chains

$$\langle M(0) \cdot M(0) \rangle = \sum_{i=1}^{N} \left\{ \sum_{j=1}^{n_{r_i}} \sum_{j'=1}^{n_{r_i}} \langle \mu_{ji}(0) \cdot \mu_{j'i}(0) \rangle \right\} \tag{4}$$

where n_{r_i} is the number of dipole groups in chain i.

If we restrict our considerations to chains containing only one type of dipolar group (as in poly(vinyl chloride), poly(ethylene terephthalate), poly(n-alkyl meth-

acrylates), polar-non-polar copolymers such as acrylonitrile-butadiene copolymers)
then

$$\langle M(0) \cdot M(0) \rangle = \sum_{i=1}^{N} \left\{ n_{r_i} \mu^2 + 2 \sum_{j=2}^{n_{r_i}} \sum_{j'=1}^{j-1} \langle \mu_{ji}(0) \cdot \mu_{j'i}(0) \rangle \right\}$$ (5)

We note that the μ_{ji} in Eq. (5) are "liquid" dipole moments and are related to
the "vacuum" dipole moment by (Fröhlich, 1958)

$$\mu_{ji} = (\mu_{ji})_{vac} \frac{(\epsilon_\infty + 2)(2\epsilon_0 + 1)}{3(2\epsilon_0 + \epsilon_\infty)}$$ (6)

For the special case of flexible chains the terms $\langle \mu_{ji}(0) \cdot \mu_{j'i}(0) \rangle$ decrease in
magnitude with increasing separation of the dipoles, so for such a system Eq. (5) is
approximated by the relation

$$\langle M(0) \cdot M(0) \rangle = V c_r \left\{ \mu^2 + \sum_{\substack{k' \\ k' \neq k}} \langle \mu_k(0) \cdot \mu_{k'}(0) \rangle \right\}$$ (7)

c_r is the number of dipoles per unit volume and the dipoles k and k' are contained in
the same chain. Thus $(\epsilon_0 - \epsilon_\infty)$ is related to auto-correlation terms $c_r \mu^2$ and pair
cross-correlation terms $c_r \langle \mu_k(0) \cdot \mu_{k'}(0) \rangle$ where the latter terms arise due to the
angular correlations of groups caused by valence angle and internal rotation factors
of the chain. As emphasized above Eqs. of the form Eqs. (3)–(7) are well-known for
dipolar media generally (Fröhlich, 1958, Kirkwood, 1939; Böttcher, 1973, 1978) and for
for polymers in particular (Fuoss and Kirkwood, 1941; McCrum et al., 1967; Cook
et al, 1970). The deductions of $\langle M(0) \cdot M(0) \rangle / V$ for chains of different chemical
structure, and in some cases of different tacticity, are given in the texts of Volkenstein (1963), Flory (1969) and Birshtein and Ptitsyn (1966). It is not always obvious
from the usual derivations how important each cross-correlation term is in relation
to the auto-correlation term in Eq. (7). As one example, we shall quote the deduction of the cross-correlation terms for model polyethers $+(CH_2)_{p-1} - O+_n$ as given
by Cook, Watts and Williams (1970) following the method developed by Read (1965).
        Following Read (1965) we write I_i as the unit vector along the i'th C–O bond
in the chain direction. The instantaneous dipole moment $P(0)$ of a chain containing n
oxygen atoms is given by (Read, 1965; Cook, Watts and Williams, 1970; Cook, 1971)

$$\langle P(0) \cdot P(0) \rangle = \mu_{co}^2 \left[\langle \sum_{n=1}^{n} (I_{np-1} - I_{np}) \cdot \sum_{n=1}^{n} (I_{np-1} - I_{np}) \rangle \right]$$

$$= \mu_{co}^2 \left[\sum_{i=p-1}^{np-1} \sum_{j=p-1}^{np-1} \langle I_i \cdot I_j \rangle + \sum_{i=p-1}^{np-1} \sum_{j=p-1}^{np-1} \langle I_{i+1} \cdot I_{j+1} \rangle \right.$$ (8)

$$\left. - \sum_{i=p-1}^{np-1} \sum_{j=p-1}^{np-1} \langle I_i \cdot I_{j+1} \rangle - \sum_{i=p-1}^{np-1} \sum_{j=p-1}^{np-1} \langle I_{j+1} \cdot I_i \rangle \right]$$

μ_{co} is the dipole moment of the C–O bond.

Each of the four terms in Eq. (8) may be expanded. For example

$$\sum_{i=p-1}^{np-1} \sum_{j=p-1}^{np-1} \langle I_i \cdot I_j \rangle = \langle I_{p-1} \cdot I_{p-1} \rangle + \langle I_{p-1} \cdot I_{2p-1} \rangle + \ldots + \langle I_{p-1} \cdot I_{np-1} \rangle$$

$$+ \langle I_{2p-1} \cdot I_{p-1} \rangle + \langle I_{2p-1} \cdot I_{2p-1} \rangle + \ldots + \langle I_{2p-1} \cdot I_{np-1} \rangle$$

$$+ \langle I_{np-i} \cdot I_{p-1} \rangle + \langle I_{np-i} \cdot I_{2p-1} \rangle + \ldots + \langle I_{np-i} \cdot I_{np-1} \rangle$$

$$(9)$$

We use the same assumptions as Read (1965): (i) all valence angles are taken to be equal to $(\pi - \alpha)$; (ii) all internal rotations of the bonds are assumed to be independent of each other; (iii) the characteristics of the internal rotation of C–O and C–C are taken to be equivalent; (iv) the potential energy function for internal rotation is taken to be symmetrical with respect to the $\phi = 0$ (i.e. trans) conformation giving $\langle \sin \phi \rangle = 0$. With these assumptions

$$\langle I_i \cdot I_j \rangle = \bar{A}_{33}^{|j-i|} \tag{10}$$

the subscript 33 denotes the 33 element, and the matrix $\bar{A}$ is given by

$$\bar{A} = \begin{bmatrix} -\eta \cos \alpha & 0 & \eta \sin \alpha \\ 0 & -\eta & 0 \\ \sin \alpha & 0 & \cos \alpha \end{bmatrix} \tag{11}$$

$\eta = \langle \cos \phi \rangle$. Eqs. (9) – (11) yield

$$\sum_{i=p-1}^{np-1} \sum_{j=p-1}^{np-1} \langle I_i \cdot I_j \rangle = n + 2(n-1) \bar{A}_{33}^{p} + 2(n-2) \bar{A}_{33}^{2p}$$

$$+ \cdots + 2 \bar{A}_{33}^{(n-1)p} \tag{12}$$

The three other terms in Eq. (8) may be similarly expanded and hence Eq. (8) becomes

$$\langle P(0) \cdot P(0) \rangle = n \mu^2 \left[1 + \left(\frac{n-1}{n} \right) \left(\frac{2\bar{A}_{33}^{p} - \bar{A}_{33}^{p+1} - \bar{A}_{33}^{p-1}}{1 - \bar{A}_{33}} \right) \right.$$

$$\left. + \left(\frac{n-2}{n} \right) \cdot \left(\frac{2\bar{A}_{33}^{p} - \bar{A}_{33}^{2p+1} - \bar{A}_{33}^{2p-1}}{1 - \bar{A}_{33}} \right) + \cdots + \frac{1}{n} \left(\frac{2\bar{A}_{33}^{2p} - \bar{A}_{33}^{(n-1)p+1}}{1 - \bar{A}_{33}} \right) \right] \tag{13}$$

$\mu^2 = 2 \mu_{co}^2 (1 - \bar{A}_{33})$ where μ is the ether group-dipole moment. It is readily shown that the coefficients of $(n - m)/n$ in Eq. (13) arise due to the orientational correlations between a reference group-dipole k say and both of its neighbouring group-

Table 1. Values of the coefficients C of $(n - m)/n$ for poly(ethylene oxide) with $\cos \alpha = (1/3)$ and for chosen values of $\eta = \langle \cos \phi \rangle$

η	$C_{(n-1)/n}$	$C_{n-2)/n}$	$C_{(n-3)/n}$
−0.5	−0.50	0.062	0.024
−0.25	−0.183	0.013	0.000
0	−0.074	−0.003	0.000
0.25	−0.187	−0.007	−0.004
0.5	−0.537	0.113	−0.038
0.75	−1.137	0.630	−0.361

dipoles at $k \pm m$ along the chain. Cook and co-workers (1970) calculated these coefficients for $p = 2, 3, 4$ and 5 for different values of n and with $\cos \alpha = \bar{A}_{33} = (1/3)$, the tetrahedral valence angle condition. In this way the magnitude of the cross-correlation functions could be estimated for different model polyethers with internal rotation restrictions. As one example of the calculations Table 1 shows values of the coefficients for various values of η for the case $p = 3$, i.e. poly(ethylene oxide). For equal probabilities of trans and gauche states $\eta = 0$. This also coincides with the condition where the internal rotation potential energy function $U(\phi)$ is a constant ("free-rotation"). Inspection of Table 1 shows that for $\eta = 0$ all cross-correlation coefficients are small compared with the autocorrelation coefficient ($= 1$). Thus the dielectric relaxation behaviour of such a model chain is essentially determined by the autocorrelation term. For $0.5 < \eta < 0$ and $0 < \eta < 1$ the chains favour gauche and trans conformations respectively. For both ranges the cross-correlation terms become increasingly important with increase in $|\eta|$. For example for $\eta = 0.75$ the cross-correlation terms are comparable with the autocorrelation term and alternate in sign. For such a chain the relaxation is to be regarded as the superposition of positive (m even) and negative (m odd) contributions.

The object of the present discussion for model chains is to emphasize that the static permittivity of a bulk amorphous polymer necessarily involves both auto- and cross-correlation terms and hence the α- and β-relaxations also involve these terms. Models for relaxation should recognise and accommodate such cross-correlation terms.

The frequency dependent permittivity $\epsilon^*(\omega)$ of a dipolar medium may be expressed by the Fourier transform relation (Hill and co-workers, 1969; Williams, 1972a 1972c; Cook and co-workers 1970; Williams, 1978)

$$\left(\frac{\epsilon^*(\omega) - \epsilon_\infty}{\epsilon_0 - \epsilon_\infty} \right) p(\omega) = 1 - i\omega \int_0^\infty \Lambda(t) \exp(-i\omega t) dt \tag{14}$$

$$\Lambda(t) = \frac{\langle M(0) \cdot M(t) \rangle}{\langle M(0) \cdot M(0) \rangle} \tag{15}$$

ϵ_0 and ϵ_∞ are the limiting low and high-frequency permittivities respectively, $p(\omega)$ is an internal field factor and $\Lambda(t)$ is a normalized time-correlation function. $\langle M(0) \cdot M(t) \rangle$ is a generalization of the equilibrium mean square dipole moment

[Eq. (5)] to the dynamic situation and may be written as (Cook and co-workers, 1970)

$$\langle M(O) \cdot M(t)\rangle = \sum_{i=1}^{N} \left\{ \sum_{j=1}^{n_{ri}} \sum_{j'=1}^{n_{ri}} \langle \mu_{ji}(O) \cdot \mu_{j'i}(t)\rangle \right\} \tag{16}$$

For polymers containing only one type of dipole Eq. (16) becomes

$$\langle M(O) \cdot M(t)\rangle = \sum_{i=1}^{N} \left\{ \sum_{j=1}^{n_{ri}} \langle \mu_{ji}(O) \cdot \mu_{ji}(t)\rangle + 2 \sum_{j=2}^{n_{ri}} \sum_{j'=1}^{j-1} \langle \mu_{ji}(O) \cdot \mu_{j'i}(t)\rangle \right\} \tag{17}$$

For flexible-chain polymers of high molecular weight Eq. (17) is approximated by the relation

$$\langle M(O) \cdot M(t)\rangle = Vc_r \left\{ \langle \mu_k(O) \cdot \mu_k(t)\rangle + \sum_{k \neq k} \langle \mu_k(O) \cdot \mu_{k'}(t)\rangle \right\} \tag{18}$$

Thus Eq. (15) becomes, for the conditions which lead to Eq. (18)

$$\Lambda(t) = \frac{\langle \mu_k(O) \cdot \mu_k(t)\rangle + \sum\limits_{k' \neq k} \langle \mu_k(O) \cdot \mu_{k'}(t)\rangle}{\mu^2 + \sum\limits_{k' \neq k} \langle \mu_k(O) \cdot \mu_{k'}(O)\rangle} \tag{19}$$

$$\equiv g(t)/g(O)$$

$g(t)$ is a generalization of the Kirkwood correlation factor $g(O)$ (Kirkwood, 1939; Fröhlich, 1958) to the dynamic situation. $\langle \mu_k(O) \cdot \mu_k(t)\rangle$ is the time auto-correlation function for the reorientation of a representative dipole k. The decay of this term from μ^2 to zero may lead to contributions to the α-, β- and $(\alpha\beta)$-processes. $\langle \mu_k(O) \cdot \mu_{k'}(t)\rangle$ are the time cross-correlation functions for the relative motion of dipoles k and k' contained in the same chain. These terms may relax from $\langle \mu_k(O) \cdot \mu_{k'}(O)\rangle$ to zero and may contribute to the α-, β- and $(\alpha\beta)$-processes. Above T_g all compexions are available to the chains and Eqs. (3) – (7) and (14) – (19) express the equilibrium and dynamic behaviour, respectively, of the amorphous polymer. In the glassy state these equations will not apply since the chains, being effectively immobile, will not achieve their equilibrium distribution of conformations. Thus in the glass only a part of $\langle M(O) \cdot M(O)\rangle$ and of $\langle M(O) \cdot M(t)\rangle$ contribute to ϵ_0 and $\epsilon^*(\omega)$ respectively. This is further discussed by Williams and Watts (1971c).

Whilst the equilibrium theory outlined above is well-known, the time-correlation function representation is not so well-known. Time-correlation functions have been used frequently in the recent literature for a variety of relaxation, spectroscopic and scattering phenomena for liquids and solids. The analysis of Kerr-effect relaxation (Beevers and co-workers, 1976), fluorescence depolarization (Valeur and

Monnerie, 1976; Berne and Pecora, 1976), quasi-elastic scattering of laser radiation
(Berne and Pecora, 1976), nuclear magnetic resonance (Abragam, 1961), quasi-elastic
neutron scattering (Allen and Higgins, 1973), vibration-rotation spectra (Rothschild et al.,
1975) and dielectric relaxation (Hill and co-workers, 1969; Böttcher, 1973, 1978;
Williams, 1972 a, 1975, 1978) data for liquids have involved various time-correlation functions for translation and rotation of molecules. Also models for molecular
motion and computer simulations of the motions of assemblies of molecules are conveniently represented or processed using time-correlation functions. Reviews of this
work are now available and for introductory accounts the reader is referred to Zwanzig (1965) and Williams (1972 a, 1978). We emphasize that Eqs. (3) − (7) and
(14) − (19) are only the starting points for our understanding of the static and dynamic permittivity data for a given polymer. Whilst the time-correlation relations,
Eqs. (14) − (19), clearly state the relations between macroscopic and molecular
quantities, measurements of $\epsilon^*(\omega)$ for a given polymer will not (a) allow a separation of the data into the component auto- and cross-correlation terms, (b) give the
mechanisms for the decay of auto- and cross-correlation terms. Additional and complementary evidence is required from other experiments in order to do this. This
will be discussed in Section II D b below.

D. Models for Molecular Motion

a. General Considerations

It seems highly desirable to give a rationalization of the occurrence of α-, β- and
$(\alpha\beta)$-processes without specifying detailed models for motion. This has been partly
described in earlier publications (Williams, 1972b, c; Williams and Watts, 1971a;
Williams et al., 1972a; Williams and Watts, 1971c).

We assume that a representative dipole k may find itself in a wide variety of
temporary local environments at the arbitrary time $t = 0$. It is assumed that as time
develops the dipole is partially relaxed via the local motions in a particular environment r (k) say (i.e. for dipole k) and is characterized by a relaxation function
$\varphi_{\beta r(k)}(t)$. The dipole is subsequently totally relaxed by an α-process which involves
the collapse of the local environment, and is characterized by a relaxation function
$\varphi_{\alpha(kk)}(t)$. It follows that

$$\langle \mu_k(0) \cdot \mu_k(t) \rangle = \mu^2 \, \varphi_{\alpha(kk)}(t) \left[\sum_{r(k)} {}^{\circ}p_{r(k)} \, q_{\alpha r(k)} + \sum_{r(k)} {}^{\circ}p_{r(k)} \, q_{\beta r(k)}(t) \, \varphi_{\beta r(k)}(t) \right]$$

$$\equiv \mu^2 \, \varphi_{\alpha(kk)}(t) \left[\mathscr{A}_{kk} + \mathscr{B}_{kk} \, \psi_{\beta(kk)}(t) \right] \qquad (20)$$

$q_{\alpha r(k)} + q_{\beta r(k)} = 1; q_{\alpha r(k)} = [\langle \mu_k \rangle_{r(k)}]^2/\mu^2$ and is the fraction of μ^2 which is not
relaxed by the $\beta_{r(k)}$ process. ${}^{\circ}p_{r(k)}$ is the equilibrium probability of obtaining the
environment r (k) for the dipole k. The individual processes for the various environ-

ments sum to give an effective strength factor $\mathscr{B}_{kk}$ for the β-process and $\mathscr{A}_{kk}$ for the α-process; $\mathscr{A}_{kk} + \mathscr{B}_{kk} = 1$. The effective decay function for the overall β-process is $\psi_{\beta(kk)}(t)$. Similarly the cross-correlation function between dipoles k and k' may be written as

$$\langle \mu_k(0) \cdot \mu_{k'}(t) \rangle = \langle \mu_k(0) \cdot \mu_{k'}(0) \rangle \varphi_{\alpha(kk')}(t) [\mathscr{A}_{kk'} + \mathscr{B}_{kk'} \psi_{\beta(kk')}(t)] \quad (21)$$

$\mathscr{A}_{kk'} + \mathscr{B}_{kk'} = 1$ and $\mathscr{B}_{kk'}$ is the fraction of $\langle \mu_k(0) \cdot \mu_{k'}(0) \rangle$ which is relaxed by local processes. These are the same local processes that gave rise to the partial relaxation of the auto-correlation function. The long-time decay of the cross-correlation function occurs via the collapse of the local environments around dipoles k and k' giving an α-process characterized by $\varphi_{\alpha(kk')}(t)$ where we would expect $\varphi_{\alpha(kk')}(t) \approx \varphi_{\alpha(kk)}(t)$, for all k and k'. Thus the total correlation function is given by

$$\Lambda(t) = \left\{ \mu^2 \mathscr{A}_{kk} \varphi_{\alpha(kk)}(t) + \sum_{k' \neq k} \langle \mu_k(0) \cdot \mu_{k'}(0) \rangle \mathscr{A}_{kk'} \varphi_{\alpha(kk')}(t) \right\}$$

$$+ \left\{ \mu^2 \mathscr{B}_{kk} \varphi_{\alpha(kk)}(t) \psi_{\beta(kk)}(t) + \sum_{k \neq k} \langle \mu_k(0) \cdot \mu_{k'}(0) \rangle \mathscr{B}_{kk'} \varphi_{\alpha(kk')}(t) \psi_{\beta(kk')}(t) \right\}$$

$$\times \left[\mu^2 + \sum_{k \neq k} \langle \mu_k(0) \cdot \mu_{k'}(0) \rangle \right]^{-1} \quad (22)$$

From Eq. (22) we have the following:
1) If the local motions occur at a much faster rate than the microbrownian motions (α-process) then $\varphi_{\alpha(kk)}(t) \psi_{\beta(kk)}(t)$ and $\varphi_{\alpha(kk')}(t) \psi_{\beta(kk')}(t)$ decay as $\psi_{\beta(kk)}(t)$ and $\psi_{\beta(kk')}(t)$ respectively and hence $\Lambda(t)$ decays in two stages giving rise to α- and β-processes of relative magnitude

$$\frac{\Delta \epsilon_\beta}{\Delta \epsilon_\alpha} = \frac{\mu^2 \mathscr{B}_{kk} + \sum_{k'} \langle \mu_k(0) \cdot \mu_{k'}(0) \rangle \mathscr{B}_{kk'}}{\mu^2 \mathscr{A}_{kk} + \sum_{k'} \langle \mu_k(0) \cdot \mu_{k'}(0) \rangle \mathscr{A}_{kk'}} \quad (23)$$

The β-process involves auto- and cross-correlation terms and since it is assumed to be a weighted sum of elementary processes occurring in a variety of local environments, it may be rather broad in the frequency domain. The α-process is also a weighted sum of processes, and relaxes the portion of $\langle M(0) \cdot M(t) \rangle$ which remains after the β-process has occurred. It is emphasized that although the α- and β-processes have different mechanisms for motion, the two processes are inter-related since the same dipoles contribute to both. This leads to a conservation relation $\Delta \epsilon = \Delta \epsilon_\alpha + \Delta \epsilon_\beta$.
2) If the α-process becomes so slow that it cannot be followed in an accessible time-scale, then only the β-process will be observed experimentally. The system will be in a non-equilibrium state, the glassy state ($T < T_g$). Equations (3) – (7) and (14) – (23) must be modified to accommodate the situation that $\varphi_{\alpha(kk)}(t)$ and $\varphi_{\alpha(kk')}(t)$ do not decay. We obtain (Williams and Watts, 1971a; Williams, 1972b).

$$(\epsilon_0 - \epsilon_\infty) = \frac{4\pi c_r \mu^2}{3kT} \left\{ \frac{3\epsilon_0 (2\epsilon_0 + \epsilon_\infty)}{(2\epsilon_0 + 1)^2} \right\} \left\{ \mathscr{B}_{kk} + \sum_{k'} \mathscr{B}_{kk'} \right\} \tag{24}$$

$$\left(\frac{\epsilon^*(\omega) - \epsilon_{\infty\beta}}{\epsilon_0 - \epsilon_{\infty\beta}} \right) p(\omega) = 1 - i\omega \mathscr{J} \left[\frac{\mathscr{B}_{kk} \, \psi_{\beta(kk)}(t) + \sum_{k'} \mathscr{B}_{kk'} \, \psi_{\beta(kk')}(t)}{\mathscr{B}_{kk} + \sum_{k'} \mathscr{B}_{kk'}} \right] \tag{25}$$

$\mathscr{J}$ indicates the one-sided Fourier transform (see Eq. (14)). Equations (24) and (25) indicate that only a portion of $\langle M(0) \cdot M(0) \rangle$ is relaxed for $T < T_g$. Also Eqs. (22) and (25) indicate that the β-process is continuous from above to below T_g, in accord with experimental observations (Williams, 1966b; Williams, 1972b; Johari, 1972, 1973).

3) If the sample temperature for $T > T_g$ is increased, the decay functions for α- and β-processes become comparable in a certain temperature range. If the temperature dependence of the α-process is greater than that of the β-process in this temperature region and if the α- and β-processes have independent mechanisms, then at higher temperatures $\varphi_{\alpha(kk)}(t)$ and $\varphi_{\alpha(kk')}(t)$ would be extrapolated to decay faster than $\psi_{\beta(kk)}(t)$ and $\psi_{\beta(kk')}(t)$ so $\Lambda(t)$, Eq. 22, decays via the α-processes giving

$$\Lambda(t) = \frac{\mu^2 \, \varphi_{\alpha(kk)}(t) + \sum_{k'} \langle \mu_k(0) \cdot \mu_{k'}(0) \rangle \, \varphi_{\alpha(kk')}(t)}{\mu^2 + \sum_{k'} \langle \mu_k(0) \cdot \mu_{k'}(0) \rangle} \tag{26}$$

Thus the coalesced process at high temperatures, which is the $(\alpha\beta)$-process, relaxes all of $\langle M(0) \cdot M(0) \rangle$ and is the α-process extrapolated to higher temperatures.

This general approach to the multiple relaxations does not involve specific mechanisms for α-, β- or $(\alpha\beta)$-relaxations. It is of course possible that the specific mechanisms will differ for different systems. For example the β-process in the alkyl methacrylate polymers will involve the partial reorientations of side-group and backbone units, while for poly(vinyl chloride) and poly(ethylene terephthalate) only partial reorientations of the backbone contribute to the β-process. The experimental data for amorphous polymers appear to be consistent with this general approach. i.e.
(i) the pattern of α-, β- and $(\alpha\beta)$-relaxations is common to many amorphous polymers, as discussed above, and to small-molecule glass-forming systems; (ii) a conservation rule $\Delta\epsilon = \Delta\epsilon_\alpha + \Delta\epsilon_\beta$ is found to hold, at least semi-quantitatively, for poly-(ethyl methacrylate) (Williams, 1966b; Sasabe and Saito, 1968) and poly(n-butyl methacrylate) (Williams and Edwards, 1966; Sasabe and Saito, 1968), where the decrease in $\Delta\epsilon_\beta$ with increasing pressure is accompanied by an increase in $\Delta\epsilon_\alpha$ (see e.g. Fig. 11 of Sasabe and Saito, 1968). Interestingly the compensation rule applies to poly(methyl methacrylate)s of different stereo-regularities. The root mean-square dipole moment per repeating unit of isotactic polymer ($\mu_{eff} = 1.4$ D[4]) and syndio-

---

4 In SI-units: 1 D = 3.335 x 10^{-30} Cm.

tactic polymer (μ_{eff} = 1.27 D) (Pohl and co-workers, 1960) are not very different. Experimentally $\Delta\epsilon_\alpha \gg \Delta\epsilon_\beta$ for the isotactic polymer whereas $\Delta\epsilon_\alpha \ll \Delta\epsilon_\beta$ for the syndiotactic polymer (Shindo and co-workers, 1969; McCrum et al., 1967). Thus for the isotactic polymer the limited motions (β-process) relax only a small part of $\langle M(0) \cdot M(0) \rangle$ leaving the remainder to be relaxed by the α-process. For the syndiotactic polymer the converse is true;

   (iii) it is predicted that decrease in temperature or increase in pressure would decrease $\Delta\epsilon_\beta$ since limited motions are reduced (a) by a decrease in local volume and (b) via the Boltzmann factor — where non-equivalent local states are involved as would be the case for the side-group motion in poly(alkyl methacrylates). Such behaviour is observed for poly(alkyl methacrylates), poly(vinyl chloride) and poly-(ethylene terephthalate).

b. Specific Models

We have seen that the dielectric relaxation of amorphous polymers is conveniently considered in terms of equilibrium and dynamic auto- and cross-correlation factors for dipolar groups. Specific models for motion require an equation of motion for an assembly of polymer chains. Whilst it is possible to simulate, by computer, the motions of assemblies of atoms (Rahman, 1964), diatomic molecules (Berne and Harp, 1970) and even water molecules (Rahman and Stillinger, 1971) yielding various time-correlation function of the motions, such simulations are restricted to very fast motions ($t < 10^{-10}$ s) and so are inappropriate for solid polymers. It is therefore necessary to use model theories for chain motion, and for assemblies of chains, which it is hoped give a reasonable representation of the actual behaviour. Table 2 indicates some of the models used for dielectric relaxation of small molecules and for polymers[5]. The original Debye theory (Debye, 1929) for small-angle rotational diffusion in a viscous continuum was extended by DiMarzio and Bishop (1974) to the case of a viscoelastic continuum and by Ivanov (1964) and Anderson (1972) to the case of jumps of finite size. Motions in a barrier system were analyzed by Hoffman (1954, 1955) for the dielectric situation and were applied to α- and β-relaxations by Brereton and Davies (1977). Defect-diffusion theories have been developed for dielectric relaxation (Glarum, 1960; Anderson and Ullman, 1967) and viscoelastic relaxation (Phillips and co-workers, 1972a). Small-angle rotational diffusion was incorporated into a theory of the dielectric relaxation of polymer chains by Kirkwood and Fuoss (1941) and by Yamafuji and Ishida (1962) whilst stochastic models (Stockmayer 1967, 1976) and barrier models (Dubois-Violette and co-workers 1969; Geny and Monnerie, 1977; Jernigan, 1972; Beevers and Williams, 1975) for chain motions have been proposed. Note that theories which yield an average relaxation time only (e.g., free volume and related theories, see McCrum et al., 1967) are not based on an equation of motion and only reflect an average property — a transport coefficient — of the dynamical system.

---

5 Since this review is concerned with low frequency motions, Table 2 excludes models involving inertial factors. For models involving inertial and memory functions see the accounts of Berne (1971) and Williams (1978).

Table 2. Specific models for dielectric relaxation

A. *Rotational diffusion*
   Deby (1929), DiMarzio and Bishop (1974), Ivanov (1964), Anderson (1972)

B. *Barrier models*
   Hoffman (1954, 1955), Adam (1965), Brereton and Davies (1977)

C. *Defect-diffusion/fluctuations*
   Glarum (1960), Anderson and Ullman (1967), Phillips and co-workers (1972a), Beevers and co-workers (1976)

D. *Chain-dynamics*
   Kirkwood and Fuoss (1941), Yamafuji and Ishida (1962), Stockmayer (1967, 1976), Dubois-Violette and co-workers (1969), Geny and Monnerie (1977), Hayakawa and Wada (1974), Saito and co-workers (1963), Jernigan (1972), Beevers and Williams (1975)

In common with other properties arising from statistical processes, dielectric relaxation curves are very broad — being at least one decade of frequency in half-width, and as a result it is possible to fit experimental data for α- and β-processes using any of the models of Table 2 (with their adjustable parameters), this being especially true if a distribution of relaxation times (with its adjustable parameters) is introduced. Thus there has been a continuing uncertainty of the applicability of the models used in any given case. For example Williams (1964b) interpreted data for the α-relaxation of poly(methyl acrylate) using transition-state theory, small-angle diffusion theory and both the Cohen-Turnbull and Doolittle-Bueche free volume theories. This continuing problem with relaxation studies may be expressed in mathematical form as follows.

Consider the motion of a reference dipolar group k. Let the coordinates of the group-dipole be described by $\Omega = \Omega\,(\theta, \phi, \psi)$ where θ, ϕ and ψ are the Euler angles with respect to a reference coordinate system. The conditional probability of obtaining the group-dipole in the element $d\Omega = d\theta\,d\,(\cos\phi)d\psi$ around Ω at time t given the group-dipole was in the element $d\Omega_0$ around Ω_0 at $t = 0$ is given by (Berne, 1971; Berne and Pecora, 1976)

$$f\,(\Omega, t \mid \Omega_0, O)\,d\Omega d\Omega_0 =$$

$$= \sum_{J,K,M} \left[\frac{2J+1}{8\,\pi^2}\right]\, F_{J,K,M}\,(t)\,D^*_{J,K,M}\,(\Omega)\,D_{J,K,M}\,(\Omega_0)d\Omega d\Omega_0 \qquad (27)$$

The Wigner rotation matrices $D_{J,K,M}\,(\Omega)$ form an orthogonal set of functions which span the space of the Euler angles Ω. The $F_{J,K,M}\,(t)$ are time-correlation functions for the angular motion of the group. They are obtained from Eq. (27) using the orthogonality of the Wigner rotation matrices.

$$F_{J,K,M}\,(t) = \int D_{J,K,M}\,(\Omega)\,D^*_{J,K,M}\,(\Omega_0)f\,(\Omega, t \mid \Omega_0, O)\,d\,\Omega\,d\,\Omega_0 \qquad (28)$$

It is readily shown that (see Berne, 1971, p. 707)

$$\langle \boldsymbol{\mu}_k (0) \cdot \boldsymbol{\mu}_k (t) \rangle = \mu^2 \sum_K F_{1,K,0}(t)$$

$$= \mu^2 \sum_K \langle D_{1,K,0}^*[\Omega(0)] D_{1,K,0}[\Omega(t)] \rangle \qquad (29)$$

The above formulation is simplified if the group and its dipole have axial symmetry and if the motion occurs with axial symmetry with respect to its initial orientation. The distribution function may be expanded in terms of Legendre polynomials $P_n [\cos \theta (t)]$, where $\theta (t)$ is the angle between $\boldsymbol{\mu}_k (0)$ and $\boldsymbol{\mu} (t)$. We have (Berne, 1971; Berne and Pecora, 1976; Williams, 1978)

$$f(\Omega, t \,|\, \Omega_0, 0) = \frac{1}{4\pi} \sum_n (2n+1) P_n (\cos \theta) \, \xi_n (t) \qquad (30)$$

The $\xi_n (t)$ are angular time-correlation functions

$$\xi_n (t) = \int f(\Omega, t \,|\, \Omega_0, 0) P_n (\cos \theta) \, d\Omega. \qquad (31)$$

It is readily shown that (Berne, 1971; Williams, 1978)

$$\langle \boldsymbol{\mu}_k (0) \cdot \boldsymbol{\mu}_k (t) \rangle = \mu^2 \langle P_1 (\cos \theta (t)) \rangle = \mu^2 \, \xi_1 (t) \qquad (32)$$

From Eq. (29) we see that in the general case the dielectric experiment gives a sum of $F_{1,K,0}(t)$ while from Eq. (32) we see for a special case that one correlation function, $\xi_1 (t)$, is obtained. Clearly such information is inadequate to define the distribution function involved. This is the reason why so many models can be used to fit a given set of experimental data without it being possible to discriminate between them. The complete solution would be to obtain all the correlation functions by use of complementary experiments on a given system thus allowing $f(\Omega, t \,|\, \Omega_0, 0)$ to be defined and hence limiting (or constraining) the parameters of the models to fit all the data. It is not possible to devise enough experiments to do this, so the only way to make progress is to carry out as many complementary experiments as possible and then to see which of the available models can be made consistent with the data[6].

It has been demonstrated that simple small-molecule glass-forming systems exhibit dielectric α- and β-processes. It is clear from the work of Johari, Goldstein and Smyth that the interpretation of the behaviour for polymers may not require theories of chain dynamics. It appears that the behaviour of the two classes of systems are accommodated by the general approach given above [see Eq. (22)] where for small molecules only the autocorrelation functions are important while for polymers both auto- and crosscorrelation functions are involved. Williams and co-workers (Beevers and co-workers, 1976; Beevers and co-workers, 1977a, b; Crossley and Williams, 1978) studied the α-relaxation using both dielectric and Kerr-effect techniques for

---

6 A similar situation prevails for the equilibrium statistics of chain molecules. A given set of data for the mean-square dipole moment $\langle \mu^2 \rangle$ may be fitted by an arbitrary choice of internal rotation parameters. However complementary measurements of the mean-square end-to-end distance taken together with $\langle \mu^2 \rangle$ severely constrain the choice of the internal rotation parameters.

dipolar solutes in supercooled o-terphenyl, tritolyl phosphate and its mixtures with o-terphenyl and several liquid alcohols. For the simple non-associated liquids the auto-correlation function $\xi_1(t)$ and $\xi_2(t)^7$ were determined from the dielectric and Kerr-effect relaxations respectively. It was found that $\xi_1(t) \approx \xi_2(t)$ at each temperature having a form $\exp - (t/\tau_0)^\beta$ with $\bar{\beta} \approx 0.5$. Since small-angle rotational diffusion requires exponential time-functions $\xi_1(t) = \exp - 2D_R t$ and $\xi_2(t) = \exp - 6D_R t$, where D_R is a rotational diffusion coefficient, these data for the α-relaxation are inconsistent with the model of small-angle rotational diffusion. A "strong-collision" or "fluctuation-relaxation" model was proposed (Beevers and co-workers, 1976) in which the molecule only moves when it experiences a fluctuation of critical size leading to its total randomization. For this model

$$f(\Omega, t | \Omega_0, 0) = \lambda(t)\, \delta\,(\Omega - \Omega_0) + \frac{(1 - \lambda(t))}{4\pi} \tag{33}$$

δ indicates a delta function, $\lambda(t)$ is a decay function characterized by the fluctuations. From Eq. (33)

$$\langle P_n\,(\cos\theta\,(t))\rangle = \lambda(t) \tag{34}$$

Thus this simple model requires all angular time-correlation functions to be equal to $\lambda(t)$. This model accomodates the dielectric and Kerr-effect data for certain viscous molecular liquids, e.g. fluorenone in o-terphenyl (Beevers and co-workers, 1977b) and tritolyl phosphate (Beevers and co-workers, 1977a). In view of the similarities between the α-relaxations of such small-molecule systems and of amorphous polymers it seems possible

(a) that the mechanism for the α-relaxation is the same for both systems and

(b) that this process follows the fluctuation-relaxation mechanism with $\lambda(t)$ being of the form $\exp - (t/\tau_0)^\beta$ with $\bar{\beta} \approx 0.4$ to 0.6. We note that whilst (a) may be true, it is possible that other mechanisms than that envisaged by Eq. (33) could lead to the result $\xi_1(t) = \xi_2(t)$ and the particular form observed for this function.

Experimentally it is possible to study the α-relaxation for liquid-glass-forming polymers using the Kerr-effect. We have recently studied a poly(phenylmethylsiloxane)^8 (Beevers and co-workers, unpublished work) using the dielectric and Kerr-effect techniques. Figure 1 shows a representative Kerr-effect transient for a 1 ms rectangular pulse of the applied field at 211.7 K. The rise and decay functions are approximately equivalent and are far broader than the single exponential function. They are fitted quite well with the Williams-Watts function with $\bar{\beta}_{\text{rise}} = 0.41$ and $\bar{\beta}_{\text{decay}} = 0.62$. The corresponding dielectric loss-curve is fitted with $\bar{\beta} = 0.48$ at this temperature. Similar behaviour was obtained at other temperatures. Figure 2 shows $\log (f_m)_\alpha$ versus T^{-1} for dielectric and Kerr-effect data. For the dielectric studies f_m is the frequency of maximum loss factor. For the Kerr-effect studies f_m is cal-

---

7 $\xi_2(t) = \langle P_2\,[\cos\theta\,(t)]\rangle = \langle 3\cos^2\theta\,(t) - 1\rangle/2$.

8 Sample supplied by Professor A. J. Barlow of Glasgow University, it has a designated viscosity of 15600 centistoke $(= 1.56 \times 10^{-2}\,\text{m}^2\,\text{s}^{-1})$.

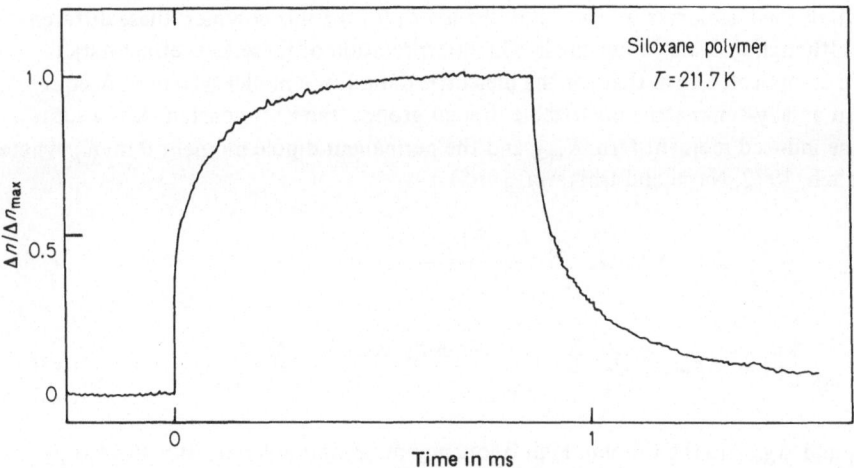

Fig. 1. Normalized electrically-induced optical birefringence against time following a rectangular pulse for a poly(phenylmethylsiloxane) at 211.7 K

culated from the transient data using the relation $\log_{10} f_m = -\log_{10} (2\pi\tau_0) - F(\bar{\beta})$ where $F(\bar{\beta}) = 0.10$ for $\bar{\beta} = 0.5$. Figure 2 shows (i) that the rise and decay Kerr-effect points differ by ≈ 0.2 (ii) that the dielectric points lie at slightly lower values of $\log f_m$ than those from the Kerr-effect transients. When it is considered that the half-

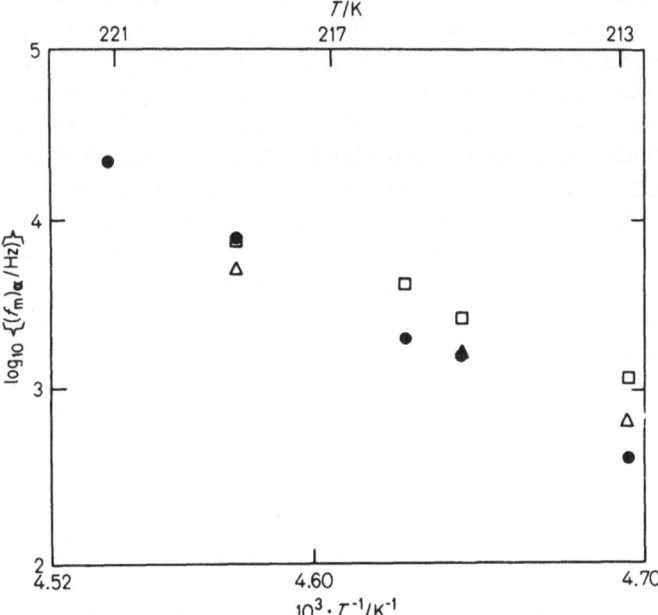

Fig. 2. Logarithm of maximum loss frequency for α-process $(f_m)_\alpha$ vs. reciprocal temperature for a poly(phenylmethylsiloxane). $\triangle$ and $\square$ refer to Kerr-effect rise and decay transients respectively; $\bullet$ are the dielectric points

width of the loss-curves is near 2.0 in the $\log f$ plot for this polymer, these differences, although real, are rather small. The interpretation of these Kerr-effect data is more complicated than that for the dielectric data. For a model system of N equivalent axially-symmetric, polarizable dipolar groups, the Kerr-constant K is a sum of the induced moment term $K_{ind.}$ and the permanent dipole moment term K_μ where (Kielich, 1972, Nagai and Ishikawa, 1965)

$$K_{ind.} = \frac{4\pi}{45\,kT}\,\Delta g_0\,\Delta g_\infty\,\sum_{i=1}^{N}\,\sum_{j=1}^{N}\,\frac{\langle\cos^2\theta_{ij}-1\rangle}{2} \tag{35}$$

$$K_\mu = \frac{4\pi}{45\,k^2\,T^2}\,\Delta g_\infty\,\mu^2\,\sum_{i=1}^{N}\,\sum_{j=1}^{N}\,\sum_{k=1}^{N}\,\frac{\langle 3\cos\theta_{ij}\cos\theta_{ik}-\cos\theta_{jk}\rangle}{2} \tag{36}$$

Δg_0 and Δg_∞ are the low and high frequency molecular polarizability anisotropy factors, θ_{ij} is the angle between the principal axes of groups i and j. Equation (35) indicates that auto-correlation terms (i = j) and P_2-type cross-correlation terms, (i≠j) contribute to $K_{ind.}$ while Eq. (36) indicates that auto-correlation terms (i = j = k), two-particle P_2-type cross-correlation terms (i ≠ j, j = k) and three-particle cross-correlation terms (i ≠ j ≠ k) contribute to K_μ. The poly(phenylmethylsiloxane) chain corresponds to a more complicated situation than the model which gives Eq. (35) and (36), so it might be supposed that the Kerr-effect data for this polymer could not be interpreted on any simple molecular basis and therefore could not be used to complement the dielectric data. Whilst this may be true for the details of the relaxation process, we may be guided by the apparently simple form of the experimental results (as was the case for certain liquid alcohols, see Crossley and Williams, 1978). The model of structural relaxation used for the α-relaxation of viscous molecular liquids (see Eqs. (33) and (34)), if applied to the more complicated polymer situation, predicts that the relaxation of all auto- and cross-correlation terms in Eqs. (35) and (36) would occur with the relaxation function $\lambda(t)$. Thus all orientational relaxation functions, including the Kerr-effect rise and decay functions and the dielectric relaxation functions would (a) have the same functional form and (b) have the same average relaxation time at a given temperature. Whilst this is not quantitatively true for the data in Figs. 1 and 2, it is certainly semiquantitatively true, so we would suggest that the α-relaxation in this polymer may be represented by the "fluctuation-relaxation" mechanism, in common with that for certain small-molecule glass-forming systems.

We now consider specific mechanisms for the dielectric β-process. For polymers containing flexible side-groups (notably the alkyl methacrylate polymers) the β-process is usually considered to be due to the motion of the side-group in a barrier system (Ishida and Yamafuji, 1961; Ishida, Matsuo and Yamafuji, 1962; Tanaka and Ishida, 1974; Shimizu, Yano and Wada, 1975). For polymers containing dipoles rigidly attached to the main chain the process is considered to be a local-mode motion, i.e., a limited damped oscillation, of parts of the chain (Yamafuji, 1960; Yamafuji and Ishida, 1962; Saito and co-workers, 1963; Hayakawa and Wada, 1974; Williams, 1966a; Williams, 1971c; McCrum et al., 1967; Wada, 1977). Our general approach, Eq. (22), would allow side-group, local-mode and other mechanisms to be included in the scheme of α-, β- and $(\alpha\beta)$-processes. One major difficulty with models for the β-process is that the equation of motion (e.g., rate equation for motion in a barrier system or diffusion equation for local mode motion of a chain) may lead to only a narrow loss curve whereas experimental curves for the β-process in most amorphous polymers are very broad with half-widths in the range 4–6 decades of frequency. Thus we have the situation, in common with that for the α-relaxation, that several models, with their adjustable parameters, may be used to fit data. Some guidance may be obtained from the results of Johari, Goldstein and Smyth (Johari and Smyth, 1969; Johari and Goldstein, 1970 a, b; Johari, 1973; see also Williams and Crossley, 1978 for a recent summary). They found that the β-process for such systems as chlorobenzene-decalin and pyridine-toluene was (a) small, with $\Delta\epsilon_\beta \ll \Delta\epsilon_\alpha$, was (b) broad with half-widths near five decades of frequency, and (c) importantly, coexisted with the α-process above T_g (see Johari, 1972; Johari, 1973). It was suggested (Johari, 1972 that the β-process arose from the non-cooperative rearrangement of a fraction of the molecules which sit in a range of local environments – leading to broad relaxations. This is as envisaged in Eq. (22) above. Given the fact that the β loss-curves are so broad it seems unlikely that we can be more specific about its mechanism for small molecule or for polymer glass-forming systems. It appears that an advance can only be made when complementary information, in the time or frequency domains, becomes available from Kerr-effect, NMR relaxation and quasi-elastic light scattering experiments.

We not that Jonscher (1975 a, b, 1977) and Brereton and Davies (1977) have given recent discussions of the origin and form of the dielectric α- and β-relaxations. Jonscher has suggested an empirical relation for $\epsilon^*(\omega)$, has discussed its limiting high frequency behaviour and has proposed a screened-charge-hopping model for relaxation. Brereton and Davies considered α- and β-relaxations in terms of a two-state model in which the energy difference W between states is assumed to be given by the relation

$$W = W_0 + k T_c (n_1 - n_2)/(n_1 + n_2) \tag{37}$$

n_1 and n_2 are the average occupation numbers of the two states. The analysis excludes the cross-correlation terms $\langle \mu_k (0) \cdot \mu_{k'}(t) \rangle$ and is confined to the behaviour of the auto-correlation term $\langle \mu_k (0) \cdot \mu_k (t) \rangle$. The model yields one process, governed by a single exponential decay, at each temperature. The behaviour above T_c is quite different from that below T_c and there is a singularity at T_c. Above T_c the process is regarded as the α-process and the model gives a log (f_m) vs. T^{-1} plot which is strongly curved in the manner of WLF behaviour. Below T_c the log f_m values first *increased,* reach a maximum and then decreased with decreasing temperature. This process is regarded as the β-process. It is important to note that this model does not allow α- and β-processes to coexist above T_c. The plots given for ϵ'' against T at fixed frequencies give two peaks, but as the authors point out, the β-process ceases to exist above a certain frequency. In the present account we have emphasized that the α-, β- and $(\alpha\beta)$-processes exist for polymers with and without flexible side groups and for small molecule glass-forming systems. The theory of Brereton and Davies does not appear to be consistent with the pattern of behaviour described in Sections IIB,

II C and II Da). We not that Johari (1972, 1973) has convincingly demonstrated that α- and β-processes coexist for certain small-molecule systems above T_g, as has been discussed above, and that the β-process is continuous in its properties through the glass-transition range.

III Crystalline Polymers

A. Introduction

Many accounts of the dielectric behaviour of partially crystalline polymers are now available (McCrum et al., 1967; Ishida, 1969; Hedvig, 1977; Baird, 1973; Wada, 1977; Williams and Crossley, 1978; Sasabe, 1971; Saito, 1964; Hoffman et al., 1966). No attempt can be made in this Section to give a comprehensive account of this and recent work. We shall consider only limited and selected aspects of the dielectric behaviour of polymers of medium-crystallinity and high-crystallinity. This account will be aimed at emphasizing molecular mechanisms for relaxation, realising of course that, as discussed above in Section II, it is essential to have complementary experimental evidence in order to establish specific relaxation mechanisms.

B. Polymers of Medium Degree of Crystallinity

It is well-known that polymers such as poly(ethylene terephthalate), poly(vinylidene fluoride) and polycarbonate may achieve a degree of crystallinity χ up to about $0.5 - 0.7$ ($50 - 70\%$ crystalline) when crystallized in bulk from the melt or from the quenched glassy polymer (Billmeyer, 1971; McCrum et al., 1967). The dielectric α- and β-processes observed for the amorphous polymer may appear in modified form in the partially crystalline polymer and new processes, due to motion in the crystalline phase, may also appear. In the following account we consider only the behaviour of poly(ethylene terephthalate) and poly(vinylidene fluoride), two polymers which have been studied recently in some detail.

On crystallizing poly(ethylene terephthalate) from the melt or from the glass (by heating a glass above T_g) the resultant α-process (i) has $(\Delta \epsilon_\alpha)_{\text{cryst.}} \ll (\Delta \epsilon_\alpha)_{\text{amorphous}}$, (ii) is much broader than that for the amorphous polymer and (iii) occurs at much lower frequencies, at a given temperature, than that for the amorphous polymer (McCrum et al., 1967; Ishida, 1969; Saito, 1964). The cause of (i) is simply that less dipoles contribute because they are immobilized in the crystalline regions. The origin of (ii) and (iii) is less certain and will be discussed below. For the β-process $(\Delta \epsilon_\beta)_{\text{cryst.}} < (\Delta \epsilon_\beta)_{\text{amorph.}}$, as expected, but the shape and frequency-temperature location of the process are essentially unaffected by the increase in χ.

Since a poly(ethylene terephthalate) sample having $\chi \approx 0.5$ appears in the optical microscope to be wholly spherulitic it is obvious that the apparently ordered spherulites contain about 50% disordered material, where this disorder is static and dynamic in its nature. Sawada and Ishida (1975) studied the dielectric α-process during the isothermal crystallization of the amorphous polymer at 95 °C. They found that ϵ_m'' and $\log f_m$ decrease continuously with a crystallization time ranging from 70 min to 4465 min, $\Delta\epsilon_\alpha$ decreased to 50% of its initial value during this time but the degree of crystallinity χ (as judged by density measurements) had only increased to 0.25 (i.e., 75% amorphous). Thus the loss process in the specimen was smaller than anticipated on the basis of the amorphous content and this was attributed by Sawada and Ishida (1975) to a restriction of motion in the "amorphous" regions due to the adjacent crystalline regions. The effective dipole moment in the disordered regions is lowered due to the constraints on motion due to the ordered regions. Similar studies of polyethylene terephthalate have been made by Tidy and Williams (1978). Figure 3 shows their data for a sample undergoing crystallization from the amorphous polymer at 106.7 °C. Successive curves are for different times of crystallization. As the crystallization proceeds the spherulites grow into the amorphous phase until they impinge, filling all the space. Figure 3 reflects this process. For $t = 0$, 20 and 40 min the single large peak is the α-process for the truly amorphous material. We see that at $t = 40$ min there is a low frequency shoulder to the main peak which grows with increasing time so that at 60 min the low and high frequency peaks are of similar magnitude. For $t > 80$ min there remains the small low-frequency process (α'-process, say) which is due to main-chain microbrownian motions within the "amorphous" regions within the spherulites. A careful analysis of this and similar data shows that $\Delta\epsilon_{\alpha'}$ first increases rapidly with increasing χ, reaches a maximum, then slowly decreases, being indicative of secondary crystallization within the crystallites at long times. Tidy and Williams (1978), in common with Sawada and Ishida (1975), found that $\Delta\epsilon_{\alpha'}$ was less than that expected on the basis of χ values and

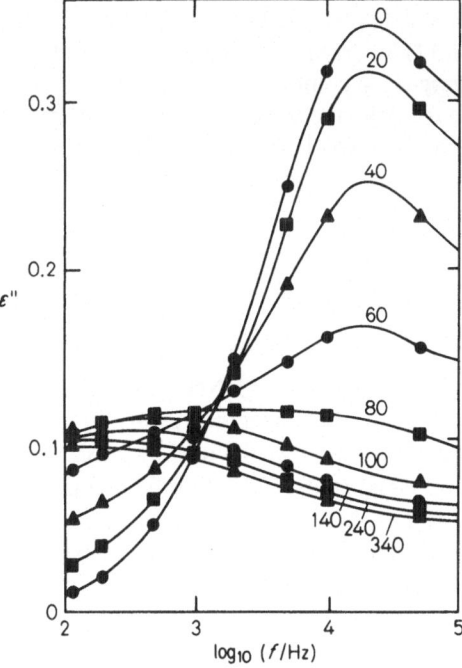

Fig. 3. Dielectric loss-factor ϵ'' against logarithm of frequency f for a poly(ethylene terephthalate) sample crystallizing at 106.7 °C. Numbers at curves indicate the times of crystallization (in min) following the rapid heating of the amorphous glass to 106.7 °C

thus it appears that the spatial freedom of the mobile groups in the disordered regions within the crystal is considerably restricted by the presence of the ordered regions. In addition it seems that the mobile chain segments find themselves in a variety of environments in the spherulite leading to an increased breadth of the α'-relaxation over that for the α-relaxation (reminiscent of the β-process). Thus two distinct processes α and α', due to motions in the normal amorphous phase and in the disordered regions within spherulites respectively, have been observed for poly(ethylene terephthalate). We anticipate that similar behaviour would be obtained for other polymers having a medium degree of crystallinity.

With regard to the β-process for poly(ethylene terephthalate) whilst one may assert that this is due to the limited motions of the chains it is difficult to establish a mechanism since the loss-curves are so broad.

One of the most extensively studied polymers of medium crystallinity is poly-(vinylidene fluoride). The interest in this polymer results from its use as a piezo-electrical and pyro-electrical material (Nakamura and Wada, 1971; Hayakawa and Wada, 1973; Murayama, 1975; Murayama and co-workers, 1975; Murayama and Hashizume, 1976; Mopsik and Broadhurst, 1975). Many dielectric studies have been made and recent work includes that of Koizumi and co-workers (1969), Sasabe and co-workers (1969), Yano (1970), Kakutani (1970), Osaki and co-workers (1971), Nakagawa and Ishida (1973), Uemura (1974), Osaki and Ishida (1974), Yano and co-workers (1974) and Brereton and co-workers (1977). It is evident from these and earlier studies that the dielectric behaviour of this polymer is one of the most complex of the linear polymer systems. Studies have been made of samples of different (i) crystal forms, (ii) orientations, (iii) crystallinity, (iv) thermal and electrical histories. At least four relaxation regions are observed, being labelled here as α_i, α_c, α_a and β in the order of decreasing temperature of appearance at a given frequency. We note that poly(vinylidene fluoride) is obtained in α-crystal form (form II) by melt crystallization and in the β-crystal form (form I) by drawing specimens. The α and β forms have helical (TGTG') and planar zigzag chain conformations respectively. Both conformations have appreciable dipole moment and thus, when the chains are mobile, large permittivities are obtained (ϵ_0 in the range 10 to 20). The α_c, α_a and β processes appear to be intrinsically dipolar in origin and will be discussed first. The α_i process appears to be due to ion-diffusion processes.

i. The α_c Process

This process has been studied by Koizumi and co-workers (1969), Sasabe and co-workers (1969), Kakutami (1970), Osaki and co-workers (1971), Nakagawa and Ishida (1973), Osaki and Ishida (1974) and by Yano and co-workers (1974). In some of these studies the low frequency part of the absorption was obscured by a rising loss at lower frequencies (α_i process). However Osaki et al. (1971) showed that this could be removed by pretreating specimens with a steady electric field (up to $25 \text{ kV} \cdot \text{cm}^{-1}$) at $160\,^\circ\text{C}$. The treatment causes a migration of charges and an electrolysis. Removal of the field results in a slow partial recovery of the low frequency process. For the α_c process itself, it is found (for the α-crystal form) that (i) ϵ''_m increases with increased χ and increased lamella thickness, (ii) $\log f_m$ decreases with increasing χ and lamella thickness, (iii) $\log f_m$ vs. T^{-1} plots are linear with $Q_{app} \approx 25$ kcal/mol[9], (iv) the ϵ'' vs. $\log f$ plots are symmetrical and have a halfwidth near 2 decades of frequency. The evidence suggests that the α_c process occurs in the crystalline phase of the polymer. Its analysis requires a model for relaxation. Nakagawa and Ishida (1973) analyzed their data using a two-site model similar to that proposed by Hoffman et al. (1966) for which coupled motions of the chains in the fold surface and the crystal interior are considered. Their results indicate that, whilst it appears that the folds are mobile and the crystal interior undergoes limited motions, the interior motions, due to the large dipole moments that are involved, make the largest contribution to α_c. Orientation, which leads to a decrease in the α-crystal form and an increase in the β-crystal form, results in a diminution of the α_c process (Koizumi and co-workers, 1969; Kakutani, 1970) but Kakutani (1970) has suggested that the α_c process in the β-crystal form occurs 50 °C higher than that for the α-crystal form. Osaki and Ishida (1974) showed that the α_c peak for a single-crystal mat specimen was about four times larger than that for the normal melt-crystallized specimen thus providing clear evidence that the α_c process is associated with the crystalline phase.

ii. The α_a Process

This process has been studied by Koizumi and co-workers (1969), Sasabe and co-workers (1969), Nakagawa and Ishida (1973), Osaki and Ishida (1974) and by Brereton and co-workers (1977). It is found (i) that ϵ''_m decreases with increasing χ, (ii) that the $\log f_m$ vs. T^{-1} plots are WLF in form and lie on essentially the same curve for different specimens, and (iii) that the ϵ'' vs. $\log f$ plots are very broad at low temperatures but narrow and increase in magnitude with increasing temperature. It is suggested, by most authors, that the α_a process is due to the gross-microbrownian motions of the chains in the amorphous parts of the polymer. Nakagawa and Ishida (1973) have analyzed the relaxation time behaviour and that for $\Delta\epsilon_{\alpha a}$ in terms of the Adam-Gibbs theory of cooperative relaxation. It was assumed that the TGTG' conformation of the α-crystal phase was also maintained locally in the amorphous regions, and from their values for $\langle\mu^2\rangle/n$ they deduced that the size of the cooperatively rearranging regions decreased with increasing temperature. We note that the α_a process for many amorphous polymers has a shape which changes only slightly with temperature whereas the α_a process for this polymer has a halfwidth $\Delta \log f$ which increases from 2 to 6 on going from -1.2 °C to -35.4 °C. Thus it is quite different from that for many amorphous polymers, and it would appear that the chain motions are strongly affected by the presence of crystallites — as was the case for poly(ethylene terephthalate).

iii. The β Process

The β process is observed as a small high-frequency shoulder to the α_a process and has $(\epsilon''_{m\beta}/\epsilon''_{m\alpha_a}) \approx 0.1$. The process is extremely broad in the frequency domain

---

9 In SI-units: 1 kcal = 4.184 kJ.

(see e.g., Sasabe and co-workers, 1969, Fig. 3) and has an apparent activation energy of about 2 kcal/mol. Its mechanism is generally assigned to be a local mode relaxation in the amorphous regions.

iv. The α_i Process

As indicated above, at the lowest frequencies and at high temperatures there is an increasing low-frequency loss, termed the α_i process (Sasabe and co-workers 1969). Uemura (1974) has shown that at high temperatures (80 °C to 160 °C) and low frequencies (10^{-2} to 10^2 Hz), samples of poly(vinylidene fluoride) exhibit very large permittivity and loss factors, with ϵ' being in the range 10 to 10^5. He found that ϵ' (f) and ϵ'' (f) followed $f^{-3/2}$ and f^{-1} laws respectively, at a given temperature. Whilst ϵ'' (f) did not exhibit maxima, $\tan \delta = \epsilon''$ $(f)/\epsilon'$ (f) exhibited distinct maxima for 160 °C and 136 °C in the range 10^{-2} to 10^2 Hz. These data were shown to be consistent with the theory of Uemura (1972, 1974) in which a dispersion of the permittivity occurs due to the diffusion of ions. His theoretical relations are

$$\epsilon_i' = \frac{2 n_0 q^2}{\sqrt{\pi} \, lk \, T} \left(\frac{D_0}{f} \right)^{3/2} \exp - [3E_d/2 + W/(2 \, \epsilon_0)]/(k\,T) \tag{38}$$

$$\epsilon_i'' = \frac{2 n_0 q^2}{k \, T} \cdot \left(\frac{D_0}{f} \right) \exp - [E_d + W/(2 \, \epsilon_0)]/(k\,T) \tag{39}$$

where n_0 and D_0 are the concentration and the diffusion coefficient of the mobile ions at a given temperature, q is the charge on an ion, l is the inter-electrode separation, E_d is the apparent activation energy for the diffusion of the ions and W is the dissociation energy of the ionic impurities. From a comparison of ϵ_i' (f, T) and ϵ_i'' (f, T) data E_d = 34 kcal/mol and W = 342 kcal/mol were obtained.

Yano and co-workers (1974) have studied the low frequency ($10^{1.5}$ to $10^{2.5}$ Hz) and high temperature (110 °C to 190 °C) behaviour of poly(vinylidene fluoride). They show that an abrupt change in properties occurs at 155 °C (the melting point) and 135 °C (the "freezing point") in the heating and cooling cycles respectively. They conclude that an interfacial polarization process occurs below T_m and a process due to electrode polarization occurs above T_m. They have suggested a two-phase model for the interfacial polarization process.

C. Polymers of High Degree of Crystallinity

Whilst many studies have been made on a variety of highly crystalline polymers, most attention has been given to the polyethers and to polyethylene (oxidized, chlorinated and ethylene/carbon monoxide copolymers). We shall give a brief summary of the results for these materials, emphasizing molecular mechanisms for the multiple processes.

The dielectric behaviour of poly(oxymethylene) and poly(ethylene oxide), both highly crystalline polymers, is well known (see McCrum et al., 1967; Ishida, 1969). There has been much interest in the relaxations occurring below room temperature in the kHz range. As indicated by Ishida (1969) a single crystal-mat sample of poly-(ethylene oxide) shows one broad absorption ($'\beta'$-process) at $-85\,^\circ$C for 12.8 kHz but on melting and melt-recrystalizing the specimen two processes, one (β) at $-80\,^\circ$C and the other (α_a) at $-35\,^\circ$C, are found. The β-process, which is common to both forms of sample was considered to be due to motions in imperfect regions of the crystal, whilst the α_a-process was considered to arise from motions in an amorphous phase between lamellae. This work was continued by Ishida and coworkers in a series of publications for poly(oxymethylene) Tanaka et al., 1970; Tanaka and Ishida, 1972; Tanaka et al., 1972). Tanaka et al. reported results for a large crystal of poly(oxy-methylene) (2 cm in diameter, 5 cm long) prepared from a solid-state polymerization of a single crystal of tetraoxymethylene. The X-ray patterns did not indicate the presence of an "amorphous" phase for this material, and, through its mode of preparation, chain-folded lamellae structures would presumably be absent, the polymer being helical in conformation but with "extended chain" structure only in the crystal. One small low temperature (β) process was observed near $-80\,^\circ$C (800 Hz) and no loss-peaks were observed at higher temperatures. The β-process decreased to (1/3) of its original magnitude on acetylation of the end-groups of the chains in the crystal. Also it was found that the β-process was anisotropic with respect to the crystallographic axes, with $(\Delta\epsilon_\perp/\Delta\epsilon_\parallel) \approx 6$ where $\perp$ and $\parallel$ mean perpendicular and parallel with respect to the long axis (C-axis) of the chain. A melt-crystallized specimen of poly-(oxymethylene) gave a process which was highly asymmetric at low frequencies, being indicative of two superposed processes, one as in the single crystal material (lower T process) and the other presumably from the amorphous regions of the melt-crystallized material (as was the case for poly(ethylene oxide) mentioned above). In a later paper Tanaka et al. (1972) showed that poly(oxymethylene) crystals prepared by a solid-state polymerization of single crystals of trioxymethylene and pentaoxy-methylene also exhibited a small low-frequency low-temperature loss process, and they found that ϵ''_m for the trioxymethylene polymer was about (1/2) of that for the tetraoxymethylene polymer and was about (1/10) of that for the pentaoxyme-thylene polymer. Ishida and co-workers concluded from these works that the β-process in the solid-state polymerized material was due to the (anisotropic) reorientation of OH end groups in defective regions within the crystal. Tanaka and Ishida (1972) calculated the potential energy contour for the reorientation of a terminal OH group and a terminal acetyl group in the crystal and found two minima in the first case and several minima in the other. This analysis indicates that the OH group may reorientate a two site barrier system, of non-equivalent sites having an energy difference ≈ 1 kcal/mol, with a barrier of ≈ 4 cal/mol. The analysis also indicates that the acetylated terminal group scarcely moves due to the high barriers opposing motion. Whilst the analysis by Tanaka and Ishida involves detailed conformational energy calculations and leads to a prediction of a single relaxation time process for the OH motion, the experimental loss curves have a half-width of around four decades of frequency thus precluding a simple discrete process. The breadth was attributed to a non-uniform molecular environment for the terminal group. There is little doubt, however, that the low temperature process in melt-crystallized poly-(oxymethylene) is a composite of at least two processes, and that the lower tempe-rature component involves the reorientation of terminal groups in the crystal.

Due to its practical importance as a cable material and due to its special position as a model polymer, polyethylene has been the subject of numerous dielectric studies. In common with poly(vinylidene fluoride), polyethylene shows a complicated relaxa-tion behaviour. Since the chain is intrinsically non-dipolar the material is oxidized, chlorinated or is made by copolymerizing carbon monoxide with ethylene, so that dipoles can be introduced to act as a probe on the motions of the parent material. Extensive accounts of the early experimental work are available (McCrum et al., 1967; Ishida, 1969; Baird, 1973; Hedvig, 1977; Wada, 1977; McCall, 1969) and Hoff-man et al. (1966) have proposed models for the mechanisms of the multiple relaxa-tions. More recently, studies have been made for oxidized polyethylenes (Kakizaki and Hideshima, 1973; Ashcraft and Boyd, 1976) for ethylene/carbon monoxide co-polymers (Phillips and co-workers, 1971; Phillips and co-workers, 1972b; Stark-weather, 1977) and for chlorinated polyethylenes (Matsuoka and co-workers, 1971; Ashcraft and Boyd, 1976). Yano and co-workers (1977, 1978) have studied the dielectric loss of high density polyethylene at temperatures below 4.2 K. Nakagawa and Tsuru (1976) have discussed the reduction of loss in radiation-polymerized poly-ethylene. It is not possible to do justice to these careful and comprehensive studies in this review, rather we shall summarize some of the findings.

Phillips and co-workers (1971, 1972b) observed α-, β- and γ-processes for copoly-mers of ethylene and carbon monoxide which contained 0.5−1% carbon monoxide comonomer. Their orientation studies indicated that the α-relaxation was due to motion in the crystalline regions. They showed that $\Delta\epsilon_\alpha$ could be used to study the orientation function of the crystals in oriented specimens. Their results were in ac-cord with those of Davies and Ward (1969) for oriented lightly oxidized polyethylene. We note that Davies and Ward found that the β-process did not exhibit anisotropy indicating that this process was due to motion in the inter-crystalline regions. Kaki-zaki and Hideshima (1973) found α-, β- and γ-processes for oxidized polyethylene where the γ-process was a composite process. Matsuoka and co-workers (1971) studied linear and branched polyethylenes and linear polyethylene chlorinated to concentra-tions ranging from 0.1 to 4 moles per cent. For a 0.49 mol-% specimen α- and γ-processes were observed. On increasing the chlorine content to 1.76 mol-% and 3.19 mol-% both α- and γ-processes increased in magnitude and the β-process appeared. A solution crystallized sample of the 1.76 mol-% material showed only α- and γ-processes with losses being $\approx (1/10)$ of those for the melt crystallized material. Ash-craft and Boyd (1976) have described their extensive studies of the α-, β- and γ-relaxa-tions for oxidized and chlorinated linear and branched polyethylenes. This study is perhaps the most comprehensive to date and we can only summarize their results

(i) The α-process in oxidized linear polyethylene disappears discontinuously on melting and reappears on cooling in a manner expected from the crystallization be-haviour. Thus the α-process is associated with the crystalline regions. The relaxation strength $\Delta\epsilon_\alpha$ falls below that expected for full dipole participation indicating a selec-tive partitioning of CO groups that favours the amorphous regions. $\log (f_m)_\alpha$ de-creased with increasing lamella thickness for various polyethylenes and levels off approximately, at large lamella thicknesses. Such behaviour is consistent with a chain rotation involving twisting at long chain-lengths — as discussed by Fröhlich (1958), Hoffman et al. (1966) and Williams et al. (1967).

(ii) The β-process was found for the oxidized branched polymer and for the chlorinated linear polymer, and, in common with earlier workers, it was thought to be similar to a glass-rubber relaxation, i.e., due to cooperative motions in disordered regions of the polymer.

(iii) The γ-process was observed as a very broad process in all samples and was interpreted in terms of a nearly symmetrical relaxation time spectrum. The dependence of $\Delta\epsilon_\gamma$ on χ suggested that the process is an amorphous one. No evidence of a change of shape of the γ-process with χ that could be interpreted in terms of a crystalline component, in addition to the amorphous one, was found. It appeared that the γ-process was due to an (unspecified) localized conformational motion which occurs in different local packing environments (cf., the β-process discussed above for wholly amorphous polymers).

Yano and co-workers (1977, 1978) studied the dielectric behaviour of low-density and high-density polyethylenes and copolymers of ethylene and vinyl alcohol in the range 1.5 to 4.2 K. It was suggested that the small loss process occurring near 1 kHz at 1.67 K for single-crystal mat high-density polymer (which is nearly a single relaxation time process) is due to the phonon-assisted tunneling of the protons of hydroxyl groups which are attached to tertiary carbons. The orientation studies (Yano and co-workers, 1978) indicate that the protons move parallel to the C axis in the crystal.

With regard to the origins of α-, β- and γ-processes in polyethylenes it appears that there is general agreement that the α- and β-processes arise from motions associated with the crystalline phase ($\alpha \equiv \alpha_c$-process) and a disordered phase ($\beta \equiv \alpha_a$-process). There continues to be uncertainty regarding the γ-process since although all the evidence suggests that it is due to local motions of chains, it is still not well-established if these motions occur in both the crystalline and wholly disordered regions or the latter only. We note that the surface of lamellae are directly connected to the crystal interior so it is difficult to regard the lamellae surfaces as wholly disordered regions which are independent of the crystal. The fact that the γ-process is so broad in the frequency domain also makes it difficult to determine the mechanism for motion (in common with the "β_a" absorption of amorphous and partially crystalline polymers as discussed above).

The mechanism for the α-relaxation in oxidised and chlorinated polyethylenes has been considered in detail by many workers (see Hoffman et al., 1966 for several proposed mechanisms and a review of the earlier literature). In a recent paper Mansfield and Boyd (1978) have reviewed the earlier work of Fröhlich, Tuijnman, Booij, Williams and co-workers and of Boyd and co-workers on the theory of the rotation of a polymethylene chain within its own crystal. It is well-known that the molecular reorientation may occur as a rigid-rod for short chains, but for long chains the reorientation process involves a twisting of the chain — as indicated first by Fröhlich. Hoffman and co-workers (1966) and Williams and co-workers (1967) extended the Fröhlich model to include a double-parabola energy and a cosine energy function. It was found that at short chain lengths the chains would reorientate between two states as rigid rods whilst at a critical chain length, and beyond, chain-twisting was involved. Whilst Hoffman and co-workers (1966) and Williams and co-workers (1967) recognised that the reorientation involved both rotational and translational displacements of the chain in the crystal, the model did not explicitly take the detailed chain structure and the local crystal structure into account. Such detailed calculations have now been made by Mansfield and Boyd (1978). From conformation energy calculations in the crystal potential they find that reorientation may be accomplished by means of a twisted region which propagates smoothly along the chain and hence across a crystal. The twisted region is localized to about 12 CH_2 units. The twisted region differs from the point-defect proposed by Reneker in that the chain distortion is relatively uniform through the twist and that there is no shortening of the chain. The theory gives, as one result, a plot of $(Q_{app})_v$ as a function of lamella thickness l in which $(Q_{app})_v$ increases linearly with log l up to $l \approx 40$ Å and then levels off to a plateau[10]. A comparison of the experimental data for paraffins and polyethylenes (containing carbonyl groups) with the theory gives a favourable result, especially in view of the relative absence of adjustable parameters in the theoretical part.

IV Concluding Remarks

As indicated earlier, we are aware that the present account has not been comprehensive. It has not considered the interesting behaviour of polymer-diluent systems (Adachi et al., 1975; Adachi and Ishida, 1976; Adachi et al, 1977) for which multiple relaxations are observed. Also no account has been given of recent work for ion-containing polymers or for polyelectrolytes. Much work has been reported for the thermally stimulated currents for a variety of polymers (see e.g., van Turnhout, 1975, 1978). While it is clear that this technique provides a direct and rapid means of documenting multiple low-frequency processes it does not give sufficient information about the specific molecular mechanisms for individual processes. In this respect the measurement of $\epsilon^*(\omega, T)$ or $\epsilon(t, T)$ over wide ranges is to be preferred.

As mentioned above, there is a developing interest in the piezo-electrical and pyroelectrical relaxation behaviour of polymer electrets. It was clear from the early work of Furukawa and Fukada (1969) for uniaxially drawn poly(D-propylene oxide) films that piezoelectrical relaxation was related to electrical and mechanical relaxation behaviour. Subsequent studies of the complex piezoelectric coefficient and complex electrostriction coefficient for oriented polymers, including polypeptides and poly(vinylidene fluoride), have been reviewed by Hayakawa and Wada (1973), Wada and Hayakawa (1976) and Wada (1977). For oriented poly(γ-benzyl glutamate glutamate)[11] Furukawa and Fukada (1976) have documented the piezoelectric d- and e-coefficients over a wide range of frequency and temperature, and have shown the close relationship between dielectric, mechanical and piezo-electrical relaxations

---

10 $(Q_{app})_v$ is the constant volume apparent activation energy.
11 Systematic IUPAC name: Poly[imino[1-(2-benzyloxycarbonylethyl)-2-oxoethylene]].

arising from the motions of side-groups. They proposed a two-phase model of the material in order to interpret the data. Such studies show that for systems such as the polypeptides the piezo-electrical relaxation is closely related to the intrinsic molecular relaxation behaviour. Poly(vinylidene fluoride) exhibits more complicated piezoelectrical behaviour whose interpretation involves a consideration of the various crystal forms, their orientation and space-charge formation in addition to molecular relaxation factors (see e.g., Hayakawa and Wada, 1973, Wada and Hayakawa, 1976, Murayama, 1975, Murayama and co-workers, 1975, 1976).

With regard to the detailed mechanisms for motion in amorphous and crystalline polymers it is clear from the discussions of Eqs. (27)–(32) that interpretations should be made using the results of as many complementary experiments that may be devised for a given system. In this way certain mechanisms may be excluded and others favoured. However, it is not easy to devise such experiments. For transparent materials the dielectric and Kerr-effect experiments are complementary. Mechanical relaxation experiments are not easy to interpret due to the lack of a suitable theory connecting macroscopic properties to time-dependent molecular factors. The interpretation of NMR data has been made difficult by the lack of an experimental frequency coverage and due to the uncertainties regarding the mechanism for nuclear relaxation for any particular polymer system. However, considerable interest is likely for ^{13}C NMR relaxation studies of solid polymers following the work of Schaefer and co-workers (1977). It is hoped that such studies will provide the complementary information that is required for our understanding of the specific mechanisms for motion which give rise to observed multiple dielectric relaxations in solid polymers.

V References

Abragam, A.: Principles of nuclear magnetism, London: Oxford U.P. 1961

Adachi, K., Fujihara, I., Ishida, Y.: J. Polym. Sci., Polym. Phys. Ed. *13*, 2155 (1975)

Adachi, K., Ishida, Y.: J. Polym. Sci., Polym. Phys. Ed. *15*, 693 (1977)

Adachi, K., Hattori, M., Ishida, Y.: J. Polym. Sci., Polym. Phys. Ed. *15*, 693 (1977)

Adam, G.: J. Chem. Phys. *43*, 662 (1965)

Allen, G., Higgins, J. S.: Rep. Progr. Phys. *36*, 1073 (1973)

Anderson, J. E., Ullman, R.: J. Chem. Phys. *47*, 2178 (1967)

Anderson, J. E., cited In: J. Chem. Soc., Faraday Symp., *6*, 1972, p. 90 of Discussion

Ashcraft, C. R., Boyd, R. H.: J. Polym. Sci., Polym. Phys. Ed. *14*, 2153 (1976)

Baird, M. E.: Electrical properties of polymeric materials. London: The Plastics Institute 1973

Beevers, M. S., Williams, G.: Adv. Mol. Relax. Proc. *7*, 237 (1975)

Beevers, M. S., Crossley, J., Garrington, D. C., Williams, G.: J. Chem. Soc., Faraday Trans. II, *72*, 1482 (1976)

Beevers, M. S., Crossley, J., Garrington, D. C., Williams, G.: J. Chem. Soc., Faraday Trans. II, *73*, 458 (1977 a)

Beevers, M. S., Crossley, J., Garrington, D. C., Williams, G.: J. Chem. Soc., Faraday Symp. *11*, 38 (1977 b)

Berne, B. J., Harp, G. D.: Adv. Chem. Phys. *17*, 63 (1970)

Berne, B. J.: In: Physical chemistry, an advanced treatise, Vol. VIII B, The liquid state, ed. Eyring, H., Jost, W., Henderson, D., New York: Academic Press 1971

Berne, B. J., Pecora, R.: Dynamic light scattering, New York: Wiley-Interscience 1976

Billmeyer, F. W.: Textbook of polymer science, Second Ed., New York: Wiley 1971

Birshtein, T. M., Ptitsyn, O. B.: Conformations of macromolecules, (Transl. from 1964 edition by S. M. Timasheff and M. J. Timasheff), New York: Wiley-Interscience 1966

Böttcher, C. J. F.: Theory of electric polarization, Second Ed., (a) Vol. I, 1973, (b) Vol. II, 1978, Amsterdam: Elsevier

Brereton, M. G., Davies, G. R.: Polymer *18*, 1764 (1977)

Brereton, M. G., Davies, G. R., Rushworth, A., Spence, J.: J. Polym. Sci., Polym. Phys. Ed. *15*, 583 (1977)

Cook, M., Watts, D. C., Williams, G.: Trans. Faraday Soc. *66*, 2503 (1970)

Cook, M., Ph. D. Thesis, Univ. of Wales 1971

Crossley, J., Williams, G.: J. Chem. Soc., Faraday Trans. II *73*, 1906 (1978)

Davies, G. R., Ward., I. M.: J. Polym. Sci., Part B *7*, 353 (1969)

Debye, P., Polar molecules, New York: Chemical Catalog 1929

DiMarzio, E. A., Bishop, H.: J. Chem. Phys. *60*, 3802 (1974)

Dubois-Violette, E., Geny, F., Monnerie, L., Parodi, O.: J. Chim. Phys. *66*, 1865 (1969)

Ferry, J. D.: Viscoelastic properties of polymers, New York: Wiley 1961

Flory, P. J.: Statistical mechanics of chain molecules, New York: Wiley-Interscience 1969

Fröhlich, H.: Theory of dielectrics, London: Oxford U.P. 1958

Fuoss, R. M., Kirkwood, J. G.: J. Am. Chem. Soc. *63*, 385 (1941)

Furukawa, T., Fukada, E.: Nature *221*, 1235 (1969)

Furukawa, T., Fukada, E.: J. Polym. Sci. *14*, 1979 (1976)

Geny, F., Monnerie, L.: J. Polym. Sci., Polym. Phys. Ed. *15*, 1 (1977)

Glarum, S. H.: J. Chem. Phys. *33*, 639 (1960)

Hayakawa, R., Wada, Y.: Adv. Polym. Sci. *11*, 1 (1973)

Hayakawa, R., Wada, Y.: J. Polym. Sci., Polym. Phys. Ed. *12*, 2119 (1974)

Hedvig, P.: Dielectric spectroscopy of polymers. Bristol: Adam Hilger 1977

Heijboer, J.: Mechanical properties and molecular structure of organic polymers. Proc. Int. Conf. Non Cryst. Solids (Delft 1964), Amsterdam: North Holland p. 231, 1965

Heijboer, J.: Mechanical Properties of Glassy Polymers Containing Saturated Rings. Central Lab., TNO, Communic. No. 435, The Netherlands, Delft 1972

Hill, N., Vaughan, W., Price, A. H., Davies, M.: Dielectric properties and molecular behaviour. New York: Van Nostrand 1969

Hoffman, J. D., Pfeiffer, H. G.: J. Chem. Phys. *22*, 132 (1954)

Hoffman, J. D.: J. Chem. Phys. *23*, 1331 (1955)

Hoffman, J. D., Williams, G., Passaglia, E.: J. Polym. Sci. Part C, *14*, 173 (1966)

Ishida, Y., Yamafuji, K.: Kolloid Z. *177*, 97 (1961)

Ishida, Y., Matsuo, M., Yamafuji, K.: Kolloid Z. *180*, 108 (1962)

Ishida, Y.: J. Polym. Sci., Part A-2, *7*, 1835 (1969)

Ivanov, E. N.: Sov. Phys., J.E.T.P. *18*, 1041 (1964)

Jernigan, R. L., in: Dielectric properties of polymers, ed. Karasz, F. E., New York: Plenum 1972, p. 99

Johari, G. P., Smyth, C. P.: J. Am. Chem. Soc. *91*, 5168 (1969)

Johari, G. P., Goldstein, M.: J. Phys. Chem. *74*, 2034 (1970a)

Johari, G. P., Goldstein, M.: J. Chem. Phys. *53*, 2372 (1970b)

Johari, G. P., Smyth, C. P.: J. Chem. Phys. *56*, 4411 (1972)

Johari, G. P., in: J. Chem. Soc., Faraday Symp. *6*, 1972, p. 42 of Discussion

Johari, G. P.: J. Chem. Phys. *58*, 1766 (1973)

Johari, G. P.: Ann. New York Acad. Sci. *279*, 117 (1976)

Jonscher, A. K.: Nature *253*, 717 (1975a)

Jonscher, A. K.: Kolloid Z. *253*, 231 (1975b)

Jonscher, A. K.: Nature *267*, 673 (1977)

Kakizaki, M., Hideshima, T.: J. Macromol. Sci., Phys. B-*8*, 367 (1973)

Kakutani, H.: J. Polym. Sci., Part A-2, *8*, 1177 (1970)

Karasz, F. E., (ed).: Dielectric properties of polymers. New York: Plenum 1972
Kielich, S., in: Dielectric and related molecular processes, ed. M. Davies, (Spec. Period. Rep.),
    London, The Chem. Soc. *1*, 293 (1972)
Kirkwood, J. G.: J. Chem. Phys. *7*, 911 (1939)
Kirkwood, J. G., Fuoss, R. M.: J. Chem. Phys. *9*, 329 (1941)
Koizumi, N., Yano, S., Tsunashima, K.: J. Polym. Sci., Part B, *7*, 59 (1969)
Koppelmann, J., Gielessen, J.: Z. Elektrochem. *65*, 689 (1961)
McCall, D. W., in: Molecular dynamics and structure of solids, ed. Carter, R. J., Rush, J. J.,
    Washington, Nat. Bur. Stand., Spec. Public. No. 301, 1969
McCrum, N. G., Read, B. E., Williams, G.: Anelastic and dielectric effects in polymeric solids,
    New York: Wiley 1967
Mansfield, M., Boyd, R. H.: J. Polym. Sci., Polym. Phys. *16*, 1227 (1978)
Matsuoka, S., Roe, R. J., Cole, H. F., in: Dielectric properties of polymers, Karasz, F. E., ed.,
    New York: Plenum 1972, p. 255
Mopsik, F. I., Broadhurst, M. G.: J. Appl. Phys. *46*, 4204 (1975)
Murayama, N.: J. Polym. Sci. Polym. Phys. Ed., *13*, 929 (1975)
Murayama, N., Oikawa, T., Katto, T., Nakamura, K.: J. Polym. Sci., Polym. Phys. Ed., *13*, 1033
    (1975)
Murayama, N., Hashizume, H.: J. Polym. Sci., Polym. Phys. Ed. *14*, 989 (1976)
Nagai, K., Ishikawa, T.: J. Chem. Phys. *43*, 4508 (1965)
Nakagawa, K., Ishida, Y.: J. Polym. Sci., Polym. Phys. Ed. *11*, 1503 (1973)
Nakagawa, K., Tsuru, S.: J. Polym. Sci., Polym. Phys. Ed. *14*, 1755 (1976)
Nakamura, K., Wada, Y.: J. Polym. Sci., Part A-2, *9*, 161 (1971)
O'Reilly, J. M.: J. Polym. Sci. *47*, 429 (1962)
Osaki, S., Uemura, S., Ishida, Y.: J. Polym. Sci., Part A-2, *9*, 585 (1971)
Osaki, S., Ishida, Y.: J. Polym. Sci., Polym. Phys. Ed. *12*, 1727 (1974)
Phillips, M. C., Barlow, A. J., Lamb, J.: Proc. Roy. Soc., A, *329*, 193 (1972a)
Phillips, P. J., Kleinheis, G., Stein, R. S.: J. Polym. Sci. Polym. Phys. Ed. *10*, 1593 (1972b)
Phillips, P. J., Wilkes, G. L., Delf, B. W. Stein, R. S.: J. Polym. Sci., Polym. Phys. Ed. *9*, 499
    (1971)
Pohl, H. A., Bacskai, R., Purcell, W. P.: J. Phys. Chem. *64*, 1701 (1960)
Rahman, A.: Phys. Rev. A, *136*, 405 (1964)
Rahman, A., Stillinger, F. H.: J. Chem. Phys. *55*, 3336 (1971)
Read, B. E.: Trans. Faraday Soc. *61*, 2140 (1965)
Rothschild, W. G., Rosasco, G. J., Livingston, R. C.: J. Chem. Phys. *62*, 1253 (1975)
Saito, N., Okano, K., Iwayanagi, S., Hideshima, T., in: Solid state physics. Seitz, F., Turnbull, D.,
    eds., New York: Academic Press, 1963, Vol. XIV, p. 343
Saito, S.: Study of molecular motions in solid polymers by the dielectric measurements. Res. of
    the Electrotechn. Lab., Tokyo, Japan, Publ. No. 648, 1964
Saito, S., Sasabe, H., Nakajima, T., Yada, K.: J. Polym. Sci., Part A-2, *6*, 1297 (1968)
Sasabe, H., Saito, S.: J. Polym. Sci., Part A-2, *6*, 1401 (1968)
Sasabe, H., Saito, S., Asahina, M., Kakutani, H.: J. Polym. Sci., Part A-2, *7*, 1405 (1969)
Sasabe, H.: Res. of the Electrotech. Lab., Tokyo, Japan, Publ. No. 721, 1971
Sawada, K., Ishida, Y.: J. Polym. Sci., Polym. Phys. Ed., *13*, 2247 (1975)
Schaefer, J., Stejskal, E. O., Buchdahl, R.: Macromolecules *10*, 384 (1977)
Scott, A. H., Scheiber, D. J., Curtis, A. J., Lauritzen, J. I., Hoffman, J. D.: J. Res. Natl. Bur.
    Stds. *66A*, 269 (1962)
Shimizu, K., Yano, O., Wada, Y.: J. Polym. Sci., Polym. Phys. Ed. *13*, 1959 (1975)
Shindo, H., Murakami, I., Yamamura, H.: J. Polym. Sci., Part A-1, *7*, 297 (1969)
Starkweather, H. W.: J. Polym. Sci., Polym. Phys. Ed., *15*, 247 (1977)
Stockmayer, W. H.: Pure Appl. Chem. *15*, 539 (1967)
Stockmayer, W. H., in: Fluides moléculaires, Balian, R. and Weill, G., ed., New York: Gordon and
    Breach 1976, p. 101
Tanaka, A., Uemura, S., Ishida, Y.: J. Polym. Sci., Part A-2, *8*, 1585 (1970)
Tanaka, A., Ishida, Y.: J. Polym. Sci., Polym. Phys. Ed., *10*, 1029 (1972)

Tanaka, A., Uemura, S., Ishida, Y.: J. Polym. Sci., Polym. Phys. Ed. *10*, 2093 (1972)

Tanaka, A., Ishida, Y.: J. Polym. Sci., Polym. Phys. Ed. *12*, 335 (1974)

Tidy, D., Williams, G., manuscript in preparation (1978)

Uemura, S.: J. Polym. Sci., Polym. Phys. Ed. *10*, 2155 (1972)

Uemura, S.: J. Polym. Sci., Polym. Phys. Ed. *12*, 1177 (1974)

Valeur, B., Monnerie, L.: J. Polym. Sci., Polym. Phys. Ed. *14*, 11, 29 (1976)

Van Turnhout, J.: Thermally stimulated discharge of polymer electrets. Amsterdam: Elsevier 1975

Van Turnhout, J.: Thermally stimulated discharge of electrets. To be publ. In: Electrets, ed. Sessler, G. M., Topics in Physics Series, Berlin, Heidelberg, New York: Springer 1978

Volkenstein, M. V.: Configurational statistics of polymeric chains. New York: Wiley-Interscience 1963

Wada, Y., Hayakawa, R.: Japan J. Appl. Phys. *15*, 2041 (1976)

Wada, Y., in: Dielectric and related molecular Processes, ed. M. Davies (Spec. Period. Report), London, The Chem. Soc. *3*, 143 (1977)

Williams, G.: Trans. Faraday Soc. *60*, 1548 (1964 a)

Williams, G.: Trans. Faraday Soc. *60*, 1556 (1964 b)

Williams, G.: Trans. Faraday Soc. *61*, 1564 (1965)

Williams, G.: Trans. Faraday Soc. *62*, 1321 (1966a)

Williams, G.: Trans. Faraday Soc. *62*, 2091 (1966b)

Williams, G., In: Molecular relaxation processes. Chem. Soc. (London) Spec. Publ. No. 20, London: The Chem. Soc., and New York: Academic Press, p. 21, 1966c

Williams, G., Edwards, D. A.: Trans. Faraday Soc. *62*, 1329 (1966)

Williams, G., Lauritzen, J. I., Hoffman, J. D.: J. Appl. Phys. *38*, 4203 (1967)

Williams, G., Watts, D. C.: Trans. Faraday Soc. *66*, 80 (1970)

Williams, G., Watts, D. C., in: NMR, basic principles and progress, Vol. IV, NMR of Polymers, Berlin, Heidelberg, New York: Springer, p. 271 1971a

Williams, G., Watts, D. C.: Trans. Faraday Soc. *67*, 2793 (1971 b)

Williams, G., Watts, D. C.: Trans. Faraday Soc. *67*, 1971 (1971 c)

Williams, G., Watts, D. C., Dev, S. B., North, A. M.: Trans. Faraday Soc. *67*, 1323 (1971)

Williams, G., Hains, P. J.: Chem. Phys. Lett. *10*, 585 (1971)

Williams, G., Cook, M., Hains, P. J.: J. Chem. Soc., Faraday Trans. II. *68*, 1045 (1972a)

Williams, G.: Chem. Rev. *72*, 55 (1972a)

Williams, G., In: J. Chem. Soc., Faraday Symp. No. 6, 1972, p. 44 of discussion (1972b)

Williams, G., in: Dielectric properties of polymers, Karasz, F. E., ed., New York: Plenum 1972, p. 17 (1972c)

Williams, G., Watts, D. C., Nottin, J. P.: J. Chem. Soc., Faraday Trans. II, *68*, 16 (1972b)

Williams, G., Hains, P. J.: J. Chem. Soc. Faraday Symp. *6*, 14 (1972)

Williams, G., in: Dielectric and related molecular processes, ed. M. Davies (Spec. Period. Report), London, The Chem. Soc. *2*, 151 (1975)

Williams, G.: Chem. Soc. Rev. *7*, 89 (1978)

Williams, G., Crossley, J.: Ann. Rep. A, London, The Chem. Soc. *11* (1978)

Yamafuji, K.: J. Phys. Soc. (Japan) *15*, 2295 (1960)

Yamafuji, K., Ishida, Y.: Kolloid Z. *183*, 15 (1962)

Yano, S.: J. Polym. Sci., Part A-2, *8*, 1057 (1970)

Yano, S., Tadano, K., Aoki, K., Koizumi, N.: J. Polym. Sci., Polym. Phys. Ed. *12*, 1875 (1974)

Yano, O., Saiki, K., Tarucha, T., Wada, Y.: J. Polym. Sci., Polym. Phys. Ed. *15*, 43 (1977)

Yano, O., Kamoshida, T., Sekiyama, S., Wada, Y.: J. Polym. Sci., Polymer Phys. Ed. *16*, 679 (1978)

Zwanzig, R.: Ann. Rev. Phys. Chem. *16*, 67 (1965)

Received March 14, 1979

W. Kern (editor)

The Nature and Application of Electrical Phenomena in Polymers

H. Block

Donnan Laboratories, University of Liverpool, P.O. Box 147, Liverpool L69 3BX, Great Britain

Electrical and dielectric behaviour of polymers reflect macromolecular structure and motion, both in solution and the solid state. Some polymers which have special electrical properties may have commerical potential. Mention need only be made of polymer electrets, pyro-electric polymers, photo-conductive polymers as used in electro-imaging, and conductive polymers to indicate the expansion of use over that of insulators. The separation of electrical behaviour into 'dielectric' and 'bulk conductive' properties is convenient and has been followed in this review.

Table of Contents

1 Introduction

It is not many years ago when the interest in the electrical properties of polymers
was effectively limited to their electrical insulating ability. This situation has changed
very markedly in recent years, and it is likely that the future will show an even greater
expansion of interest into the electrical properties of polymeric materials. The reason
is basically two fold. Firstly, electrical, and in particular dielectric behaviour are excellent diagnostic properties, in that they reflect macromolecular structure and
motion, both in solution and the solid state. Secondly, speciality polymers which
may have commercial potential as high cost, low tonnage materials are increasingly
being developed, and this group includes polymeric systems which have special electrical properties. Mention need only be made of polymer electrets, pyro-electric polymers, photoconductive polymers as used in electro-imaging, and conductive polymers
to indicate the expansion of use over that of insulators. My object in this review is to
outline some of the advances in both the scientific application of electrical properties to the understanding of macromolecular systems, and to indicate when potential
or actual use may be made of the electrical attributes of a polymer. In the latter con-
nection it should be stressed that the usefulness of a polymer with these electrical
applications frequently depends as much on the retention of desirable polymeric
properties as the introduction of special electrical ones.

There are several generalities that can be made about the electrical property of
a material including polymers. The imposition of an electrical field on a material will
cause a redistribution of any charges in the material, provided they are mobile enough
to respond in the time scale of the field application. If some of the mobile charges
are able to diffuse throughout the sample they will support a d.c. conductance, provided that charge migration through the electrode-sample interface is possible. Alternatively, if the diffusion of mobile charges under the field is spatially limited, the
material can be polarized by the field but d.c. conductance would be absent.

Limitations in charge mobility can be of two kinds. Firstly, the charge may
migrate to boundaries over which further transport is either restricted or totally
inhibited. Under this category, the boundary may delineate the molecule as is the
case for electronic polarizability, or extend further in for example the ion atmosphere of a polyelectrolyte in solution, or constitute a phase boundary semi- or
impermeable to charge in a solid material. For the last two examples the phenomenon is termed interfacial or Maxwell-Wagner polarization. The second type of restricted mobility occurs when dipoles are present which can cause polarization by a redistribution of their inclinations relative to the field direction. Such orientation polarization need not involve complete rotatory diffusion of the dipole; even a restricted
rotation or libration can affect polarization by this mechanism.

The separation of electrical behaviour into 'dielectric' and 'bulk conductive' properties is convenient and has been followed in this review. There has been some
selection in the material covered. In part this reflects the authors interest, but it is
also a consequence of the amount known about the electrical properties of polymers.
Coverage of this now large field would require several volumes, and in fact a number
of excellent texts and review articles are available which deal with aspects of the
field. The reader is referred to some of these in the appropriate parts of this review.

Not covered are certain electromagnetic effects such as light scattering, aspects of Kerr effect measurements, dynamic birefringence, refractive index properties and many aspects of the electrical phenomena associated with poly-ions in solution.

2 The Dielectric Properties of Polymers in Solution

The measurement of the permittivity (ϵ^*) of a polymer solution can be readily achieved over an extended frequency (f) range by using a.c. bridge methods (from 10^{-2} to ca. 10^7 Hz), resonance and heterodyne beat methods (ca. 10^6 to 10^8 Hz), transmission-line techniques (10^9 to 7.5×10^9 Hz), microwaves (10^{10} to 10^{11} Hz), far infra-red spectroscopy (10^{11} to 10^{12} Hz) and, if required, higher frequencies (i.r., visible, u.v., etc.). It is this extended frequency range which provides versatility in dielectric measurements.

As a parameter, ϵ^* is a complex quantity given by

$$\epsilon^* = \varepsilon_0 (\epsilon' - i\epsilon'') \tag{1}$$

where ε_0 is permittivity of free space, ϵ' the relative permittivity of the material and ϵ'' its dielectric loss; ϵ' and ϵ'' are functions of f because at selected frequencies relaxation or resonance processes occur. For the purpose of the present review, resonance processes involving as they do electronic or vibrational transitions in the u.v., visible or i.r. spectral regions will not be discussed further. Relaxation processes, which, as distinct from resonances, require thermal activation and are thus rate processes, do play an important part in the use of dielectric information in polymer science. Over the frequency range in which such a relaxation occurs, ϵ' drops in magnitude and a loss peak appears in the ϵ'' dependence (Fig. 1). In fact

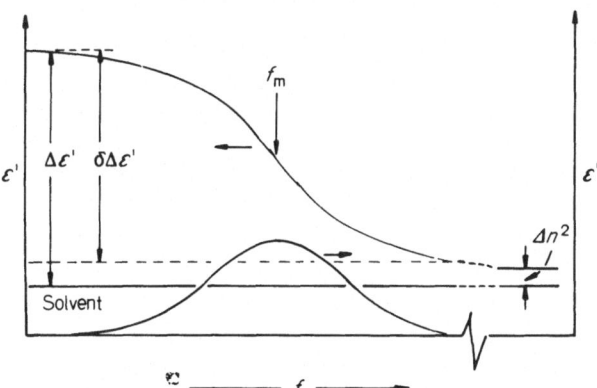

Fig. 1. The behaviour of permittivity over a single relaxation process showing the position of maximum loss (f_m) and the quantities used in estimating resultant dipole moment

ϵ' and ϵ'' are inter-related via a Fourier integral[1]. In solution, and when consideration is restricted to the polymer component, the effective part of the relative permittivity which is of interest is given by the dielectric contribution,
$\Delta\epsilon' = \epsilon'_{solution} - \epsilon'_{solvent}$. Its magnitude, or the extent to which $\Delta\epsilon'$ decreases over a relaxation ($\delta\Delta\epsilon'$), together with the frequencies over which this relaxation occurs, provides the information from which deductions at the molecular level can be made. ϵ'' equally provides such information, although as distinct from ϵ' differences are not involved since ϵ'' only has non-zero values over a relaxation, and it is presumed that the solvent itself does not relax over the relaxation frequencies for the polymer.

The magnitude of $\delta\Delta\epsilon'$ is related to the polarization of the field by the solute using the polarization mechanism (orientational or interfacial) and proceeding with a rate or rates related to the field frequency f. If f is low enough to be certain that all polarizing mechanism can operate, then $\Delta\epsilon'$ reflects the total polarization by all mechanisms. In solution the practical realization of such a low frequency presents little difficulty (but this is not always so in the solid state). A high frequency measurement can then be used to remove the electronic and atomic polarization. Often such measurements at frequencies above the relaxation regions are not attempted, and refractive index (n) measurements serve as an alternative. Here the contribution by electronic polarization to $\Delta\epsilon'$ is given by $\Delta n^2 = n^2_{solution} - n^2_{solvent}$ when n is generally measured at the Na(D)-line. Atomic polarizability makes a small further contribution (10–15% of Δn^2) which can frequently be ignored. Either method establishes the change in $\Delta\epsilon'$ due to all the dielectric relaxation mechanisms. If the polarization is due to dipole orientation, the change in $\Delta\epsilon'$ over such an extended frequency range is related to the equilibrium distribution of dipoles in space and not at all to the mechanism by which reorientation is achieved. This fact alone provides a very important method of investigating the dipole distribution and hence the chain conformation in solution. As well as this equilibrium situation, the rate aspects of the process of orientation can be studied by investigating the frequency dependence of ϵ' and ϵ'' and locating the relaxation process or processes. These equilibrium and kinetic aspects of dipolar orientation will be treated separately. Interfacial polarization will then be briefly discussed, followed by the effect of high electrostatic or flow fields on the dielectric properties of polymers in solution.

2.1 Conformation and the Effective Dipole Moment of Polymers in Solution

The dipole moment for a macromolecule in solution is composed of the vectorially summed contributions of dipoles located along the chain. This provides via dipole moment measurements, information on the configuration of chains, and in certain cases insight into the sequence distribution in copolymers[2]. Configurational analysis, and in particular the use of the rotational isomeric state model[3] provides a very direct method of estimating such resultant dipole moments as an average square $\langle \mu^2 \rangle$, as well as other chain vector parameters, such as mean-square end-to-end separations ($\langle r^2 \rangle$) or mean-square radii of gyration ($\langle s^2 \rangle$). Permittivity measurements, along with viscosity and light scattering experiments can provide configurational information for certain polymers. Such studies are very desirable in enhancing our understanding of how chain configuration is determined by the structural and energetic factors involved in the rotation of links in the polymer chain.

In a homopolymer, dipoles associated with the residues may be fixed to the chain backbone in such a way that orientation of those dipoles requires movement of the chain backbone, or they may be involved in more or less flexible side chains (Fig. 2). The former situation leads to a simpler application of statistical mechanics in estimating $\langle \mu^2 \rangle$. It should be appreciated that a dipole in a side chain usually implies that there must be a component which contributes to polarization by back-bone re-orientation, as well as components whose orientation may be achieved by local modes in the side-chain. This situation is illustrated by poly(methyl acrylate) in Fig. 2(ii). The components of dipole attached to the chain may have a contribution

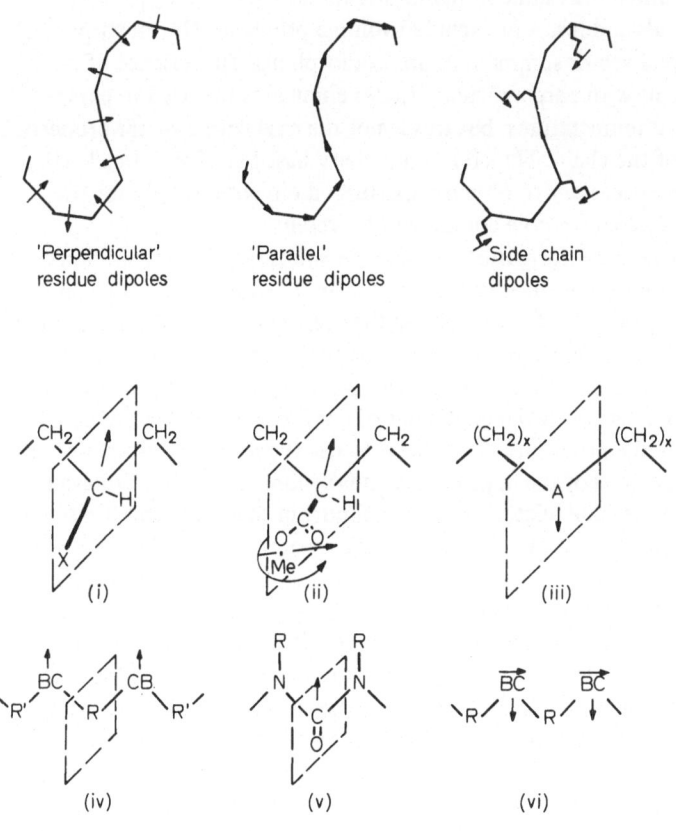

'Perpendicular' residue dipoles   'Parallel' residue dipoles   Side chain dipoles

(i)   (ii)   (iii)

(iv)   (v)   (vi)

Fig. 2. Perpendicular, parallel and side-chain residue dipoles in a polymer chain. Examples of polymer types (*i*), a vinyl polymer; (*ii*), poly(methyl acrylate) showing side-chain dipole; (*iii*), unbranched polyoxides [A=O], polyimines [A=NH], polysulphones [A=SO$_2$] etc.; (*iv*), poly-esters [B=CO, C=O], and polyamides [B=CO, C=NH] based on di-acids; (*v*), polyisocyanates; (*vi*), poly(α-ω esters) [B=CO, C=O], poly(α-ω amides) [B=CO, C=NH], poly(propylene oxide) [R=CH$_2$, B=CH·CH$_3$, C=O] etc. The planes of symmetry indicated in (*i*)–(*v*) result in the ab-sence of a parallel dipole component in such polymers, whilst the lack of a plane of symmetry in (*vi*) leads to the presence of a parallel dipole component along the chain direction ('parallel' type) as well as one bisecting a bond angle ('perpendicular' type). The preponderance of polymeric structures with a sequence of planes of reflection along the chain direction do make the latter type much more prevelant. Thus, all vinyl homopolymers, linear carbon chain poly(oxides) and poly(imines), and many others have no parallel component as shown in Fig. 2(i–iii). This figure also illustrates the presence of perpendicular and parallel components in α–ω polyesters and polyamides and the effect of branched chains in polyoxides and polysulphones. No example of a polymer with parallel but no perpendicular components is known. In one major respect the absence of a parallel component of dipole moment provides a significant advantage to permittivity measurements in chain configuration studies over techniques measuring chain dimensions. This advantage stems from the consequence that such perpendicular dipoles sum under all chain configurations irrespective of long range interaction[2, 4]. In contrast long range interactions do limit the number of valid configurations applicable either to parallel-dipoles or skeletal bonds and result in excluded volume problems. Obviously parallel-dipole and skeletal bond vector summations are isomorphous. The absence of excluded volume problems with perpendicular dipoles eliminates the need to experiment in θ solvents at θ temperatures, but this is not the case with a vector property running as the flight of the chain. Thus for permittivity based configurational estimates, θ conditions are necessary to obtain unperturbed dimensions only for the rarer situation where a parallel dipole component is present.

The relating of dielectric information on polymer solutions to configurational prediction based on structure involves two steps: the application of a suitable and efficient mathematical procedure for calculating the average resultant dipole moment of the polymer chain, and an appropriate theoretical basis for estimating this value from permittivity measurements. For most polymers the first step must involve a statistical procedure because of the large number of conformers possible[1]. Very effective, and generally the procedure of choice is to use the rotational isomeric state approximation[2, 3]. The basis of the approximation is to limit the number of available orientations of any residue relative to its neighbours in the chain, this limitation being determined by the rotamer states with potential energy minima. For example in a polymer chain from vinyl monomers the choice would be the *trans*(t) and two *gauche* (g^+, g^-) positions. The statistical weights of a limited sequence of such arrangements can then be estimated from rotamer energy differences, and other interactions such as spatial interferences can be allowed for. To illustrate from the example of an n-alkyl chain using pairs of conformers, six statistical weight terms involving tt, tg^+, tg^-, g^+g^-, g^-g^+, g^+g^+ and g^-g^- can formally be constructed. However, spatial interference in g^+g^- and g^-g^+ give rise to such a high energy state that their population is insignificant, that is the Boltzmann or statistical weight term is zero. This type of spatial interference is frequently referred to as a 'pentane type interference'.

Statistical treatments with such a local set of conformational probabilities then enables the calculation of a chain configurational partition function, or an average vector property for a chain of any length. This is most conveniently done by matrix

---

1   A notable exception are the poly-(α-amino acids) in helical form where a single conformer leads to a single structure for the chain in solution.

manipulations for which purpose a statistical weight matrix U_i for a particular skeletal bond i is defined in terms of statistical weight terms. For vector average calculations a co-ordinate transformation matrix T_i which transforms the local co-ordinates of the ith bond to the (i−1)th bond, and a bond (l_i) or dipole (m_i) vector defining the skeletal bond or dipole vector in the ith coordinate system are also required. With simple homopolymers this is sufficient for the purpose, but with copolymers, including stereocopolymers which arise in any monomer sequence where tactic placements are possible, the chain statistics cannot be developed by operation with U_i alone. Here, the copolymer sequence, even if stereoisomeric, has to be generated using Monte Carlo methods and a generator matrix G_j. The final result of such calculations provides a dipole moment ratio $\langle \mu^2 \rangle_0/(nm)^2$ (where n is the number of skeletal bonds carrying dipole moments m) or $\langle \mu^2 \rangle_0/(zm^2)$ (where z is the degree of polymerization and m the dipole moment per residue) in which the zero subscript refers to unperturbed dimensions (θ conditions) if a parallel component of dipole is present. This ratio is analogous to the bond length characteristic ratio $\langle r^2 \rangle_0/(nl^2)$ for skeletal bonds of mean square length l^2 and mean square end-to-end separation $\langle r^2 \rangle$. For chemical copolymers multiplicity of m^2 values makes the quantity $\langle \mu^2 \rangle_0/z$ more convenient.

The second step in gleaning conformational information from permittivity measurements involves the evaluation of dipole moments from experimental data. Here the major difficulty is in how to deal with the internal field problem. This problem arises from the condensed nature of the liquid state. During measurement each dipole experiences not the externally applied field but a field modified by the presence of neighbouring dipoles. In consequence the electrostatic field seen by a dipole differs from its value in the absence of the medium. There are three approaches which attempt to remedy this situation. The first is to use a solvent of such low permittivity that the sample dipoles effectively experience the external field. The Debye equation, or one of its binary component forms (such as that due to Guggenheim[5]) becomes operative. Secondly, the permittivity of the medium can be allowed for by some cavity in continuum model such as that due to Onsager[6]. The Onsager theory itself is for spherical symmetry and may be inappropriate to polymers; however, other theories of this type but with different geometrical cavities have been developed[7, 8] as has a theory which avoids the dielectric discontinuity at the cavity edge[9]. The final method is to allow for the interaction of surrounding dipoles by a correlation function treatment such as that due to Kirkwood[10] or Fröhlich[11]. Since for polymers it is not possible by dilution to eliminate intra dipole-dipole interactions, the Debye based equations are rather inappropriate for macromolecules, and the Kirkwood or Fröhlich models most appropriate. However, their application introduces difficult computations which are often also rather arbitrary. For these reasons the Onsager equations are frequently favoured. McCrum, Read and Williams[12] discuss more fully the applications of these models. Macromolecules require, as do small molecules, progressive dilution with a non-polar diluent for accurate dipole moment estimates, in order that inter-molecular dipole dipole interactions be eliminated.

Having outlined the general technique involved in using dielectric information to study conformation in solution a number of special advantages over methods which measure coil dimension should be noted. As distinct from light scattering and viscosity measurements, dipole moment estimates do not lose sensitivity for short chain molecules. Polymers, oligomers and even model short chain compounds can have their mean square dipole moments estimated within quite close limits (which arise from the internal field problem). Further, dipole measurements on short chain model compounds as a function of temperature readily provide information on the relative population of rotamers, and hence lead to estimates of rotamer state energy differences.

Quite a number of polymer structures have been analysed in terms of the rotational isomeric state model and have had comparisons made on the basis of dielectric measurements. Polyoxides of the general formula $[(CH_2)_xO]_n$ have been extensively studied with x = $2^{13-16)}$, $3^{17)}$, $4^{16, 18)}$ and $6^{19)}$, to provide information on the energetics of rotamer states which are consistent with dipole moment and characteristic ratios and their temperature dependence. Data for oligomers fit the general trend in degree of polymerization. The lack of polymer solubility limits measurements to oligomers (n = 2 and 3) for poly(oxymethylene) [x = 1] and consequently provides less extensive information for this system[20]. Possibly the greatest interest in these studies relates to deductions about the variation in rotamer preference as a function of the number of methylene groups[2]. For x = 1 there is a strong perference for the *gauche* state (7.1 kJ · mol^{-1} lower than *trans*) which results in extensive helical runs of chain, a low dipole moment ratio (≈ 0.2) and a high spatial extension for long chains. For increasing values of x the non-equivalence of main chain bonds leads to selected preferences involving g or t placements. The result is less regularity with a consequent increase in the dipole moment ratio (up to 0.6) and a decrease in the characteristic ratio for long chains. Pentane type interferences are present but are not as severe as in the polymethylene chain.

Other polymers of the perpendicular dipole type so studied include poly-(dimethylsiloxane)[21], poly(vinyl chloride)[22, 23], polystyrene[24, 25], poly(p-chlorostyrene)[26-30], and poly(p-fluorostyrene)[29]. For vinyl polymers the stereochemistry of placements effects the dipole moment ratio, and hence tacticity is reflected by dipole moment measurements. The statistics of such systems, which are stereochemical copolymers, can be developed in terms of a stereochemical replication probability p_r whose numerical value guides the probability of isotactic ($p_r \approx 0.95$), atactic ($p_r \approx 0.5$) or syndiotactic ($p_r \approx 0.05$) placement. Figure 3 shows theoretical predictions for poly(vinyl chloride) and poly(p-chlorostyrene)[23].

Comparison between theoretical calculations and experimental results are in good agreement for polystyrene and substituted polystyrenes on the basis of atactic chains. Unfortunately, the experimental information available at present on the dipole moment of poly(vinyl chloride) shows large variations between workers[22]. Nevertheless, there are good indications based on dielectric and other data that syndiotactic placements are favoured in poly(vinyl chloride). Mark[2, 23, 30] discusses the influence of side chain size in poly(vinyl chloride) and poly(p-chlorostyrene).

In all these perpendicular dipole only systems, the dipole moment ratio must become constant for sufficiently long chains. When this state is attained depends upon polymer rigidity, since the condition for constancy in dipole moment ratio is that the chain is Gaussian in the Kuhn equivalent length. That is, the degree of polymerization z must be sufficiently large so as to include the required number of 'Kuhn

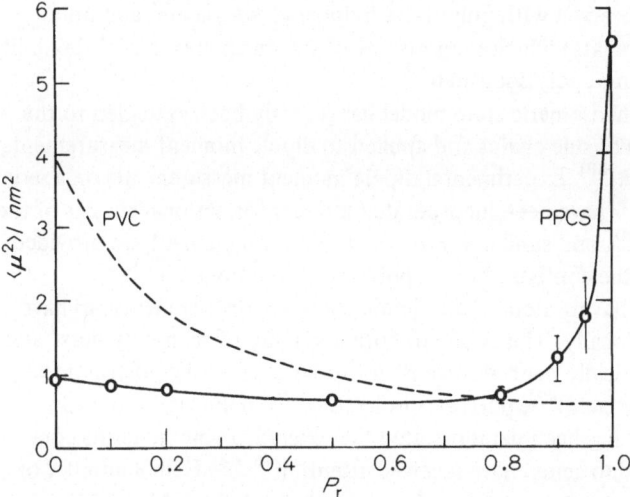

Fig. 3. Theoretical values of the dipole moment ratio at 298 K for poly(vinyl chloride) (*PVC*) and poly(*p*-chlorostyrene) (*PPCS*) chains of degree of polymerization = 100, in terms of the probability p_r of stereochemical replication (isotactic placements). Calculated points and their standard deviations have been omitted on the grounds of clarity for PVC. (Reproduced from Mark[2])

equivalent' freely linked segments. For the rather flexible polymers described above this is achieved at $z \approx 10^2$ and the resulting dipole moment ratio is rather low. A notable exception for polymers of the perpendicular dipole type are the poly(alkyl isocyanates). Here dielectric and other experimental studies[31–34] have established the presence of chains of low flexibility, a large persistence length and very high resultant dipole moments. On the basis of dielectric and other measurements on poly(n-butyl isocyanate), Tsvetkov and coworkers[33, 34] deduce that the *cis* configuration shown in Fig. 4 is strongly favoured. There is marked rigidity which results in a strong dependence of dipole ratio on z up to z ≈ 1000 for poly(n-butyl isocyanate)[33–35], poly(n-hexyl isocyanate)[35] and poly(chlorohexyl isocyanate)[36].

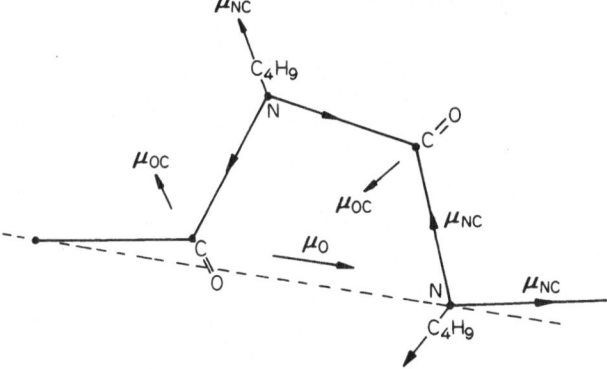

Fig. 4. The *cis* configuration of the residue in poly(n-butyl isocyanate) and the resulting dipole moment of a two unit sequence of chain in terms of bond moments (after Tsvetkov et al.[34])

Different behaviour is observed with poly(4-methylphenyl isocyanate) and poly-(4-methoxyphenyl isocyanate) which is consistent with a much greater coil flexibility for these phenyl substituted polyisocyanates[37].

The application of the isomeric state model has recently been extended to the configurational statistics of side chains and applied to dipole moment measurements in some poly(vinyl ethers)[38]. Experimental dipole moment measurements on a series of α,ω-dibromoalkanes[39] have been incorporated into conformational studies of the non polar polyethylene[40], and similarly α,ω-dihydroperfluoralkanes have provided information relevant to the configuration of poly(tetrafluorethylene)[41].

Far fewer polymers having along chain dipole components (parallel type) have been investigated dielectrically. This is due to both a sparsity of such polymers[2] and a lack of suitable dielectrically neutral solvents for some, such as the poly(amides). Poly(ω-hydroxydecanoic acid)[42], poly(ϵ-caprolactone[43]), poly(olefin sulphone)s[44] have been studied and it has been demonstrated that excluded volume terms contribute[43, 44]. Poly(α-amino acids) have received attention[45, 46]. Here a number of structures leading to unique conformers such as the α-helix are possible and provide materials of considerable dielectric interest (see Sect. 2.3).

The measurement of the dipole moments of copolymers and its analysis in terms of both sequence distribution and local chain configurations has received attention[23, 25, 47, 48]. Modern computer aided analytical procedures provide insight into the dependence of mean square dipole moment per residue on reactivity ratios, polymer composition and rotamer probabilities. One such calculation for atactic copoly-(p-chlorostyrene-p-methylstyrene)[48] has shown that at constant composition, the dipole moment is quite sensitive to the sequence distribution and thus to the reactivity ratios. This dependence would be even more marked for syndiotactic chains. For copoly(propylene-vinyl chloride)[49] and copoly(ethylene-vinyl chloride)[50] dipole moments are again very sequence dependent, much more so than the characteristic ratio. It would appear that in copolymer systems dielectric measurements can be a powerful method of investigating sequence distributions. Two copolymers, p-chlorostyrene with styrene and with p-methylstyrene have been examined experimentally[26, 51, 5]. The measurements were made on solid amorphous samples above the glass-rubber transition temperature (T_g) and they are consistent with the predictions of the rotational isomeric state model using known reactivity ratios and reasonable replication probabilities[48, 52, 53]. However, it is the view of this author that deductions based on dipole estimates for solid samples (even if amorphous and above T_g) could well be misleading, and that solution measurements are to be preferred. This is because firstly, the internal field problem is severe in the solid state and secondly, there is no guarantee that some dipoles are not very slow in orientation or effectively frozen in the time scale of the measurements. Small levels of crystallinity, local order, or just plain steric interference between chains could result in some dipoles failing to contribute to the polarization as measured. Dipole orientations in solution, even for macromolecules, are much faster than for the solid state.

---

[2] It should be appreciated that in for example polyamides and polyesters without side chain, only α, ω-amino or hydroxy acid monomers provide a non-zero along chain dipole; the symmetry of diamino- or dihydroxy-diacid condensation polymers gives zero component in such cases (Fig. 2 (iv and vi)).

2.2 The Frequency Dependence of Permittivity

It has been a condition for much of the discussion in the previous section that permittivity data refer to a frequency low enough to assume that total polarization can be achieved. However, the polarization consequent on the application of an electrical field is generally a rate process[3] involving dipole orientation or ion-migration. The study of the frequency dependence of permittivity provides insight into these processes[13, 54–56].

The polarization of the material in an electric field is consequent upon the development of an electric moment M in the material due to charge displacement under the field. Removal of the field results in the decay of $M(t)$ which can be described in terms of a decay function $\psi(t)$ or as an ensemble averaged decay of the instantaneous electric moments (permanent plus induced), $m(t)$:

$$\psi(t) = M(t)/M(O) = \langle m(O) \cdot m(t)\rangle / \langle m(O) \cdot m(O)\rangle \tag{2}$$

Further $\psi(t)$ is related to permittivity[57] by

$$1 - \frac{\epsilon^* - \epsilon_\infty}{\epsilon_0 - \epsilon_\infty} = i\,\omega \int_0^\infty dt \cdot e^{-i\,\omega t} \cdot \psi(t) \tag{3}$$

where ω is the angular frequency of the applied field. The significance of relations (2) and (3) can be highlighted by considering the orientation polarization of an assembly of dipoles $\mu(t)$ whose orientation is mutually independent and therefore decays by a first order process. The autocorrelation function can then be written as

$$\langle\mu(t) \cdot \mu(O)\rangle = \langle(O) \cdot \mu(O)\rangle e^{-t/\tau} \tag{4}$$

where τ is the relaxation time. Relations (2), (3) and (4) then lead directly to the Debye-Pellatt relation

$$\frac{\epsilon^* - \epsilon_\infty}{\epsilon_0 - \epsilon_\infty} = \frac{1}{1 + i\omega\tau} \tag{5}$$

In polymer systems such a mutually independent dipole orientation is inapplicable because dipole orientation is highly correlated. The very essence of a polymer chain generally renders independent orientation of a main chain dipole component impossible and frequently, coupling between side chain and main chain modes are involved. For a rigid chain polymer in solution, dipole orientation requires rotatory diffusion of a macromolecule as a whole, and no component due to local modes is involved, but this situation is the exception. Flexible polymers permit polarization by local mode motions as well as rotatory diffusion as illustrated in Fig. 5. Equation (5) is also inapplicable for polymers because of dispersity in molecular weight, since if the relaxation involves molecular weight dependent modes there will be a spread of relax-

---

[3] It is implied that inertial effects are not significant in the liquid or solid state.

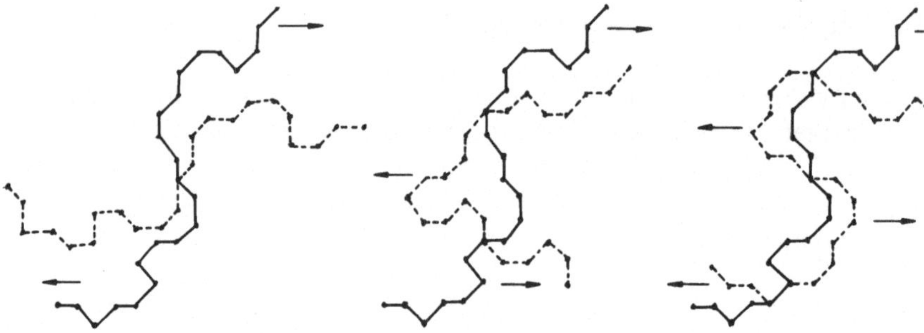

First mode   Second mode   Third mode

Fig. 5. Schematic illustration of some normal modes of motion of a polymer chain: first mode corresponds to rotational diffusion

ation times. The situation for macromolecules in solution has been discussed by Stockmayer[13] taking the normal mode analysis of Zimm[58] and Rouse[59] as a framework. Certain general deductions relating to the dielectric relaxation of polymers follow from such analysis. Dipole components attached to the chain differ in possible relaxation behaviour when they are of parallel or perpendicular type, whilst side chain components with more independent movement form a third catogory. For a parallel component, relaxation requires odd numbered modes, including the first mode (whole molecule rotation) which is often dominant in solution. The auto-correlation functions for such dipole vectors ensembly average as the chain bond vectors, and in consequence the decay function is strongly dependent on molecular weight. In such cases the relaxation frequency for maximum loss takes the form $f_m \propto M[\eta]\eta_0/T$ where M is the molecular weight, $[\eta]$ the intrinsic viscosity, η_0 the solvent viscosity and T the temperature. The constant of proportionality depends on the model [free draining[59] or non-draining[58]] and involves the orders (all odd) of modes contributing. Since $[\eta]$ is itself a function of M, the exponent of M in f_m is greater than one. For perpendicular components the situation is much more complex to model theoretically because the extent of correlation along a sequence of dipoles is reduced and depends on local structure. The Rouse-Zimm model becomes unsuitable for such local 'high-order' modes. At one extreme as pointed out by Bueche[60], for a very flexible chain f_m for such a process should be independent of M, and be governed only by the rate of change from one local conformer to another. At the other extreme, with a sufficiently rigid chain rotatary diffusion must be the relaxing mode (as pointed out by Kuhn[61]) and f_m must be M dependent. Both situations are well documented as shown by selected examples in Table 1. The expected range of behaviour is found for polymers with perpendicular dipoles. There are those whose relaxation frequency appears independent of molecular weight as well as those, such as poly(n-butyl isocyanate) whose relaxation frequency is strongly molecular weight dependant. There are examples [e.g., poly(p-chlorostyrene), poly(p-fluorostyrene), poly(N-vinylcarbazole)] where there is a change from molecular weight independent to molecular weight dependent relaxation behaviour as the molecular weight is reduced. This constitutes what must be a characteristic of all polymers of perpendicular dipole type since at sufficiently low molecular weight the whole molecule

Table 1. The dielectric relaxation of selected polymers in solution

Polymer	Solvent	T/K	f_m	$M \cdot 10^{-4}$	Dipole type/ mode[a]	Ref.
Poly(vinyl acetate)	toluene	235	5–8 MHz	1.5–200	(n + s)/Is	62)
Poly(methyl methacrylate)	toluene	263	8–13 MHz	1.4–180	(n + s)/Is	63)
Poly(butyl methacrylate)	toluene	260–266	6–10 MHz	6.1–35	(n + s)/Is	64)
Poly(vinyl bromide)	various	various	≈30 MHz	1.3–9.2	n/I	65)
Poly(p-chlorostyrene)	various	various	i) 43 MHz ii) 141–43 MHz	2–100 0.2–2	n/I n/II	29, 66, 67)
Poly(p-fluorostyrene)	benzene	298	i) 38.9–37.2 MHz ii) 97.7–38.9 MHz	2.7–15.7 0.2–2.7	n/I n/II	67)
Poly(ethylene oxide)	toluene	293	16 GHz	0.02–2.9	n/I	66, 68)
Poly(propylene oxide) (liquid)	none	253	0.63 MHz and 2–32 kHz	0.09–0.37	n/I p/III	69)
Poly(N-vinylcarbazole)	toluene	298	60–0.9 MHz	0.166–4.57	n/II to n/I	70)
Poly(hexene-1-sulphone)	benzene	298	0.12 MHz –1 kHz	70–1000	n/II to n/I	71)
Poly(n-butyl isocyanate)	various	various	1.6 MHz–30 Hz	0.32–230	n/II	31, 33)
Poly(ε-caprolactone)	1,4-dioxane	303	0.2 MHz–28 kHz and 10 GHz	3.4–8.7	p/III n/I	43)
Poly(γ-benzyl-L-glutamate)	various	various	0.1 MHz–1 kHz	2.8–50	(n + p)/III	46)

[a] *Codes:* n, normal (perpendicular); p, parallel; s, side chain. I, local segmental modes;
   Is, I cooperative with side chain; II, correlated segmental modes;
   III, whole molecule rotation relaxing dipole.

rotational mode must become dominant, whilst for sufficiently long polymer molecules a sequence of links (analogous to a Kuhn equivalent segment) will partake in
a local mode relaxation process. In this latter situation further increased molecular
weight does not influence f_m, since the relaxation involves localized motions of
equivalent segments.

It is beyond the scope of this general review to discuss in detail the various
models descriptive of locally correlated relaxation modes. Most such studies proceed by evaluating the autocorrelation function and hence $\psi(t)$. This procedure is
also of importance in analysing the solid state relaxation behaviour (Sect. 3). For
further details the reader is referred to selected papers and reviews[57, 72, 73]. What
is clear from the correlated nature of dipole relaxation in macromolecular systems
is that Debye relaxation is an unlikely process even in dilute solution. This fact has
been recognized for a number of years and was first treated by specifying a distribution of relaxation times. Semi-empirical functions descriptive of the shape of loss-
peaks abound, including that due to Cole and Cole[74], Davidson and Cole[75], Fuoss-
Kirkwood[76], Havriliak and Negami[77], Williams et al.[78] and have proved very successful in describing loss peaks in both solution and the solid state. However, it is
usually incorrect to infer that such distributions of relaxation times are real: it is
much more likely that the non-Debye character of any loss peak is the result of the
decay function not being of form $e^{-t/\tau}$, i.e., not being first order. The decay function due to Williams et al.[78] and developed on the basis of correlation functions has
recently been shown to have some theoretical basis[73].

As well as loss processes occurring over a greater frequency range than a single
Debye process, macromolecules frequently exhibit more than one relaxation process
in solution. A polymer possessing both perpendicular and parallel dipole moment
components frequently does so when the local modes responsible for perpendicular
dipoles relaxing proceeds with a higher frequency than the first order mode which
provide the major route for parallel dipole relaxation. Poly(ϵ-caprolactone)[43] and
poly(propylene oxide)[69] show examples of such behaviour. In both cases the low
frequency mode is strongly molecular weight dependent whilst the high frequency
mode is not. Measurement of ϵ' over each individual relaxation provides estimates of
resolved dipole moments which agree well with those calculated on the bases of group
moments. Similar multiplicity of peaks might arise when the polymer carries side
chain dipoles of the type shown in Fig. 2 uncoupled side chain motions could lead
to two loss processes of which that at higher frequency will be due to the spinning
mode of the side chain. There is some evidence that independent side and main chain
motions occur in poly(p-methoxystyrene)[29, 79] with the spinning of the $-$OMe group
occurring with a frequency $\geqslant 10^{10}$ Hz, but for most polymers with pendant polar
side chains such resolution is not observed (Table 1). Coupling between local main
chain and side chain modes commonly leads to a single, high frequency (10^6-10^9Hz)
process. For example the well studied polyacrylates[80], polymethacrylates[27, 63, 64, 80-82]
and poly(vinyl isobutyl ether)[27]. This tendency is not surprising since these materials all have their dipoles sited within one or two bonds from the main chain. In such
circumstances steric hinderances would require coupled side chain $-$ main chain motions
for orientational relaxation. It may be predicted that increasing the number of bonds
between the side chain and main chain could result in the appearance of well resolv-

ed, separate modes. In view of the flexibility of these chains these coupled relaxations are generally molecular weight independent in sufficiently dilute solution.

The invariance of relaxation frequency with molecular weight may not be apparent at finite concentrations of polymer, because coil — coil interactions can effect the relaxation of internal modes as well as the first order, rotational mode. There have been a number of studies of the concentration dependence of dielectric loss processes[80, 83] some of which show well the continuous trend in behaviour from dilute solution to the bulk state. Temperature variation can provide useful information on the enthalpies of activation of local mode motions. Finally, since such local modes are very structure sensitive, differences in chain tacticity would be expected, and do cause changes in loss behaviour. This prediction has been authenticated for such polymers as poly(methyl methacrylate)[82] and poly(ethyl acrylate)[27].

Dispersions in the relative permittivity of macromolecules in solution are not inevitably the consequence of the orientation of dipoles: a time dependent polarization can occur by charge migration within the framework of the polymer coil. Certainly such a phenomenon requires mobile charges to be present and is generally associated with polyelectrolyte solutions. In its simplest form, the polarization is consequent upon the field induced deformation of the counter-ion atmosphere present in the vicinity of any macroion. In terms of spheres of influence of a macromolecular phase and solvent phase on ionic conductance, any differences will lead to a Maxwell-Wagner-Sillars type interfacial relaxation[84]. Actual polyelectrolyte behaviour is frequently much more complex[54]. Multiple relaxations can occur and although in much early work the simple view of one process being due to ion-migration and others (usually one) being due to orientation polarization was put forward, it is becoming evident that ionic-migration can result in at least two loss peaks. Recent studies of ionized poly(acrylic acid)[85, 86], poly(methacrylic acid)[85, 87, 88, 89] and poly(styrenesulphonic acid)[85, 86] suggest that the two commonly observed relaxations in the kHz and MHz regions both involve charge displacement. The low frequency processes have frequency maxima which are chain lenght dependent and intensities ($\delta \Delta \epsilon'$) which depend strongly on the charge of the counter ion. Divalent ions reduce the loss magnitude. In contrast the high frequency processes have loss frequencies independent of molecular weight and are not markedly sensitive to counterion nature. It has been suggested[85, 86, 88, 89] that the high frequency process is due to Maxwell-Wagner-Sillars polarization involving polymer occluded and free ions migrating at unequal rates, thereby causing charge build up across the coil-to-free solvent interfacial region. The low frequency process is believed to involve migration from site to site of bound counterions along the ionized polymer backbone. A molecular weight dependence results from the extent of sites available for migration; specificity of ions, particularly divalent ions, is caused by tighter binding or even chelation making a hopping process more difficult.

The various mechanisms of relaxation in polyelectrolytes have been developed theoretically by a number of workers[89, 90] and there are many experimental studies on the dielectric relaxation of polyelectrolytes and biopolymers. The reader is referred to a selection of review articles on the subject [54-56, 91]. It appears that at present no clear diagnostic procedure is available for unequivocally allocating mechanisms. The influence of solvent nature, types of counter ion and their concentrations

is frequently helpful in indicating some particular mechanism. Kerr effect measurements are also useful in the study of polarization mechanism in polyelectrolytes and
biopolymers[92].

Both conventional dielectric measurements and the Kerr effect are experimental
techniques which present problems in aqueous systems due to the presence of a
large d.c. conductance (particularly in the presence of added salts). The recent observation of a forced resonance between shear and electrical fields described in
Sect. 2.3 may in future provide another technique suitable for aqueous systems.

2.3 Dielectric Measurements on Polymers in Solution when Subjected to High Electric or Shear Fields

It has been indicated in previous sections how measurements of permittivity relate
to the distribution of dipoles in a solution of macromolecules. Further, the rate processes leading to fluctuations in the spatial distribution of dipoles or positional distribution of charges are amenable to dielectric investigation. All these properties
reflect the orientation and conformational characteristics of polymer chains in
solution, and will continue to do so even if external forces alter the orientation or
topology of the polymer coil. Dielectric measurements have been found to be particularly useful in investigating the influence on macromolecular coils in solution of
high electrical fields, or shear fields in flow. High electrical field measurements are
naturally most appropriate for polymers with large summed vector dipoles or for
polyelectrolytes. In this article we shall restrict discussion of electrical field effects
to 'macro-dipoles' resulting from chain rigidity [examples being poly(α-amino acids),
cellulose derivatives and polyisocyanates] and also to changes in permittivity (dielectric saturation) rather than in refractive index (Kerr effects). For flow fields,
permittivity changes can be observed with both rigid and flexible macromolecules
and for polymer dispersions; here again the concomitant refractive index changes
(flow birefringence) will not be discussed.

The influence of an electrical field E on the angular distribution of dipoles in
solution can most simply be quantified in terms of Debye's theory for orientation
polarization. For most low field or small dipole moment situations the electrostatic energy $\mu E \ll kT$ and higher order terms in the expansion of the Langevin
function are negligible, so that the Debye equation results. For sufficiently large
E or μ the inequality is not maintained. The consequence is that under these circumstances the angular distribution of dipoles becomes significantly perturbed when a
low frequency or static field is applied. An excess of dipoles over the non-perturbed
state contribute to the polarization in the field direction and a deficiency in a direction normal to the field. If permittivity measurements are made on the dielectric in this
polarized state a reduction in ϵ' is observed because ϵ' reflects the ability for dipoles to further polarize the dielectric. The phenomenon is termed dielectric saturation and requires,
for small molecules, fields of considerable magnitude ($> 10^8$ V $\cdot$ m^{-1}). With certain
polymers having macro-dipoles much lower fields are required for marked effects to
occur ($\mu E/(kT) = 1$ when $T = 300$ K, $\mu = 4.1 \times 10^{-27}$ C $\cdot$ m or 1200 Debye and
$E = 10^6$ V $\cdot$ m^{-1}) as reported by Block and Hayes[8, 93] and Davies et al.[94]. Figure 6

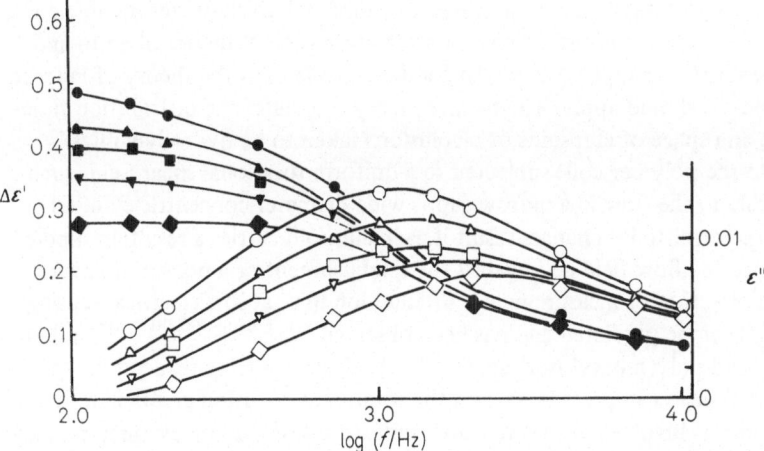

Fig. 6. The dielectric saturation of poly(γ-benzyl-L-glutamate) in solution (after Block and Hayes[8]). Field strengths in $kV \cdot cm^{-1}$ o, 0; △, 2.2; □, 3.3; ▽, 4.6; ◇, 7.2. Solid points refer to $\Delta\epsilon'$ and open points to ϵ''

shows as example the behaviour of a solution of the α-helical poly(γ-benzyl-L-gluta-mate)(PBLG) with an average dipole moment of $\approx 10^{-26} C \cdot m$. These investigations were undertaken under a static saturating field onto which a small a.c. component was added to act as sensing field. Saturation is shown by a decrease of $\Delta\epsilon'$ at low frequency plus a shift in the position of the loss peak to higher frequencies with increasing static field. Thus, the restraining influence of E is reflected in both the alignment of dipoles and their motion. The latter aspect has been analysed by Ullmann[95]. Pulsed saturating fields have been used to investigate both saturation and its decay for this polymer[96], poly(n-hexyl isocyanate)[96] and lecithin in hexane[97] where polyelectrolyte behaviour is observed. Ethyl cellulose has also been reported[8] to exhibit a ready saturation, but as expected the flexible poly(methyl methacrylate) does not show measurable saturation at fields up to $10^6 V \cdot m^{-1}$.

Just as high electrical fields can influence the spatial distribution of dipoles, so shear fields may do so when these dipoles are attached to a macromolecule. However, such studies have been few in relation to other phenomena where flow fields have an influence, namely non-Newtonian viscosity, viscoelasticity, and shear bire-fringence. In principle, dielectric measurements of this type promise certain advantages. Permittivity in reflecting the summed dipole of a macromolecule and its spatial orientation should reflect coil alignment and coil distortion for polymers in flow; loss behaviour will mirror the influence of shear on the modes of motion, and measurements at rather high shear rates (G) are relatively simple. This last advantage accrues in using Couette dielectric/shear cells in which the gap over which the shear gradient is established can be very small, much narrower than is possible for streaming birefringence studies where a parallel beam of light has to traverse through the gap. Although there have been a few early studies on the effect of shear on dielectric properties[99–102] it is only recently that equipment with the full potential of achieving significant data at shear rates which may exceed $10^5 s^{-1}$ has been described[98]. Unfortunately, the theoretical behaviour of a macromolecule in a flowing solution is not well established for all cases. For rigid, relatively undeformable coils, the theoretical work of Barisas[102] based on the more general theory of Saito and Kato[103] appears to be applicable, whilst for deformable coils the theory of Peterlin and Reinhold[104] should apply. The former theory calculates the distribution function in time and space of ellipsoids of revolution (taken to be hydrodynamically equivalent to the polymer coil) subjected to a uniform rotational shear field (such as is applicable to the flow in a narrow gap — wide diameter concentric cylinder Couette cell). Permittivity changes result if each ellipsoid carries a resultant dipole moment, since the flow field acts against the establishment of a polarization across the gap (the direction of measurement). A reduction in ϵ' at low electrical sensing frequency (f) can be predicted and has been observed for PBLG[100, 102, 105], ethyl cellulose[105] and poly(n-butyl isocyanate)[106]. Analysis of such data according to the theory of Barisas can provide data on the rotational diffusion coefficient, the dipole component involved in that rotation, indications of the coil asymmetry, and information on aspects of polymer interaction.

Figure 7 reproduces some results obtained for a sample of ethyl cellulose in benzene which show how the low frequency polarization varies with G at different substrate concentrations; its dependence extrapolated to infinite dilution, and the theoretical behaviour for differing axial ratios of coil (r) and rotatory diffusion coefficients (D). The technique can resolve differing relaxation modes in terms of their

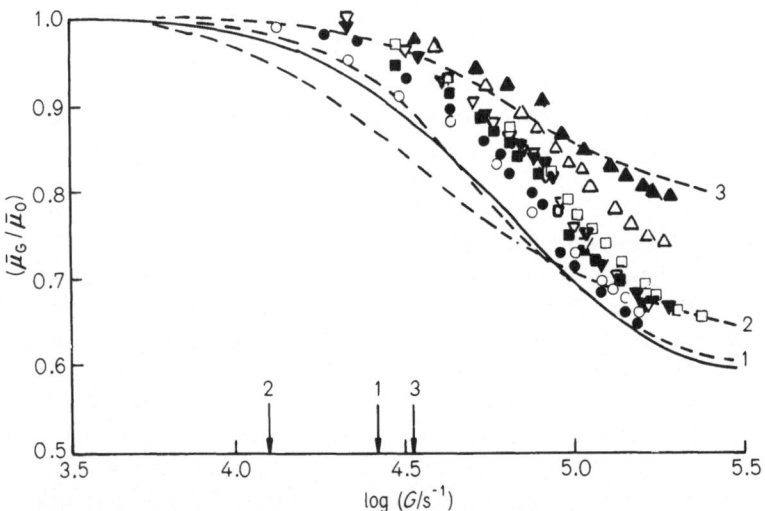

Fig. 7. The influence of flow on the polarization of the ethyl cellulose molecule in benzene after Block et al.[105]. Data show the variation of polarization per macromolecule relative to its value at zero shear rate ($\bar{\mu}_G/\bar{\mu}_0$) as a function of shear rate (G) and concentration (○, 3.7; ●, 5.6; □, 7.05; ■, 7.75; ▽, 10.6; ▼, 10.95; △, 25.2; ▲, 50 mg · cm^{-3}). The full line is the extrapolated data to zero concentration, and curves 1 and 2 correspond to the results of the application of the theory due to Barisas[102] for 40% of the total dipole relaxing and assuming spherical ($r = 1$) and needle ($r = \infty$) molecular shape respectively. Curve 3 is calculated on the basis of a 20% dipole contribution in concentrated solution and spherical symmetry. *Arrows* correspond to where $G = 2D$ thereby locating each curve

dipole contribution as shown in Fig. 7, where the effect accounts for only 40% of the total polarization. This fraction is the rotational mode contribution. In contrast the total dipole of PBLG is involved during shear-saturation[105].

The dielectric response to a shear field of more flexible polymer coils has also been recently investigated[107]. For polystyrene, poly(p-ethoxystyrene), poly(methyl methacrylate) and poly(N-vinylcarbazole) there is an increase in $\Delta\epsilon'$ with G rather than the decrease described above. The magnitude of the change in ϵ' is smaller than is the case for the more rigid polymers with larger dipoles and therefore experiments require rather higher concentrations of polymer. That an increase in ϵ' should occur for flexible polymers with perpendicular dipoles is predicted by the theory of Peterlin and Reinhold[104] who consider the change in mean residue dipole moment caused

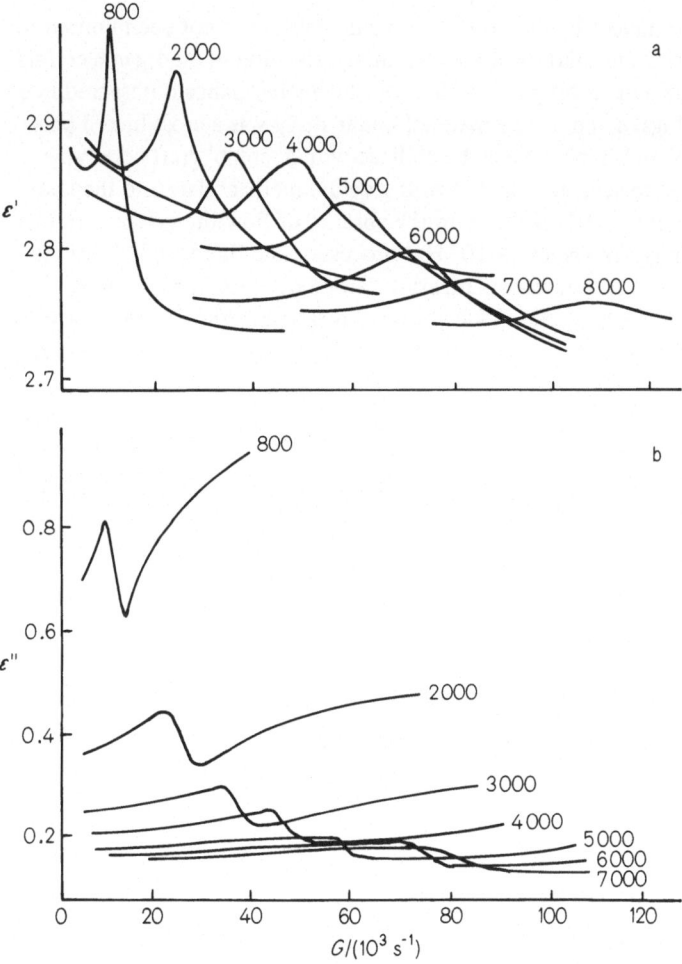

Fig. 8a–b. The behaviour of an organic colloid under shear (rate G) showing resonance in permittivity (a) and dielectric loss (b) (after Block et al.[108]). The numbers labelling individual curves refer to the electrical sensing frequency f in Hz

by flow in the vicinity of a bead and spring analogue for a polymer chain. The observations by both Block et al.[107] and Wendish[101, 104] do not support the behaviour as predicted by Peterlin and Reinhold. The complexity of the behaviour has been illustrated by Block et al. who observed a surprisingly large effect for polystyrene with its very small perpendicular residue dipole. Further, in the study of the influence of concentration, molecular weight, solvent and solution viscosity, solvent permittivity, and residue dipole on the effect, these workers obtained an empirical relation for $\Delta\epsilon'(G)$ which was independent of the residue dipole moment for polystyrene vis-a-vis its p-ethoxy substituted form. The implication that was drawn was that the phenomenon may, at least in part, reflect changes of internal field consequent upon coil distortion in flow. Since solvent molecules are involved within the environs of the coil, the solvent permittivity of these molecules becomes G dependent via an internal field change with shear, and hence solvent permittivity is a G variable term within the effect.

The study of the dielectric properties of flowing liquids has not been limited to isolated macromolecules in solution. Recently an investigation of an organic colloid system[108] has produced a novel phenomenon in which a resonance is observed in ϵ' and ϵ'' as shown in Figs. 8 and 9. The material under study was a cross-linked poly-(ethyl acrylate) dispersed in heptane and stabilized with a 'comb' graft copolymer [poly(12-hydroxystearic acid) terminated with glycidyl methacrylate and then co-polymerized with methyl methacrylate]. In the absence of flow the system exhibits low ($<10^2$ Hz) and high frequency ($\approx 10^9$ Hz) processes, and there are indications that the former is due to interfacial polarization caused by trace impurity ions. A coupling between the mechanically induced forced rotation of the spherical particles and the electrically induced drift of ions to the particle to solvent interface would

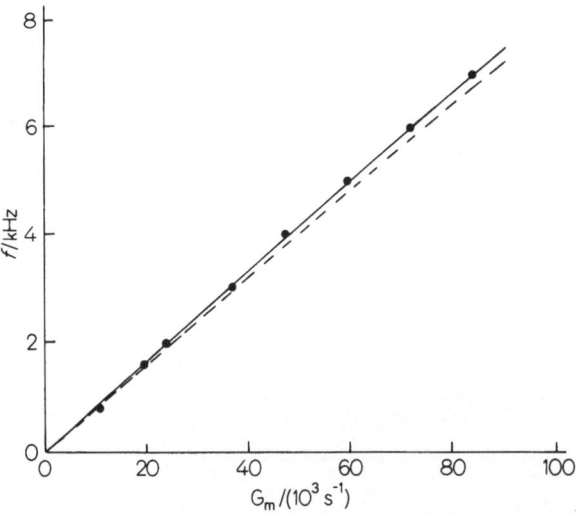

Fig. 9. The dependence of resonance position (G_m) on f for the data shown in Fig. 8 (after Block et al.[108]). The theoretical line of slope $(4\pi)^{-1}$ is shown dashed; divergence between it and the experimental points are within the limits of error in shear gap measurement

result in the polarization and orientation to be in phase at selected f and G values. The conclusion can be drawn that the resonance maxima in Fig. 8 at G values of G_{max} are proportional to the sensing frequency f and that in the absence of rotatory diffusion $G_{max} = 4\pi f$. The observations confirm this prediction (Fig. 9). As the investigators point out [108], the results promise to provide a technique for studying polyelectrolytes in conductive media (a system difficult to investigate conventionally at low frequency, p. 107–108) since the magnitude and shape of the resonance peak is presumably related to the conductance within, and rotation of, the particles. However a quantitative theory of this phenomenon is not to hand at present.

3 The Dielectric Properties of Polymers in the Solid State

The investigation of the dielectric properties of loss and permittivity has become a well established and frequently used method of probing molecular motions in macromolecules in the solid state. The field is extensive and there are a number of excellent reviews [12, 56, 109]. For these reasons this article presents only a brief and rather general view of the topic of dielectric spectroscopy in solid polymer systems, concentrating subsequently in some more detail on phenomena involving high electrical field effects.

As in macromolecular solutions, ϵ^* and its changes with frequency results for polymers in the solid state from charge movement which can be dipolar orientation or charge migration. However, there are some major differences in scale, complexity and interpretation in the solid state. Firstly, molecular processes tend to be much slower so that the frequency scale of most interest is lower than for solution work. Secondly, individual modes of motion involving dipole-orientation (and possibly charge migration if coupled to such modes) frequently occur at quite different rates when sited in differing environments. Even in the amorphous state back-bone modes can relax independently to produce broad or well separated loss processes. Thirdly, it is very difficult to relate the magnitude (in ϵ' or ϵ'') of such processes to a relaxing dipole moment. The last problem stems from both the difficulty of assigning an appropriate internal field correction to a condensed phase, and more significantly, the number of dipoles per unit volume relaxing and their angular freedom for orientation for some individual process are not generally known a priori. It would be quite incorrect to apply the Debye or Onsager theories since these assume a knowledge of the number density of relaxing dipoles and their orientability. The situation is even more ill-defined than that described in Sect. 2.1 where the total permittivity increment due to all relaxing mechanisms is presumed known. For these reasons, intensity data are less significant than information on the position and breadth of loss processes in the solid state, a deduction amply confirmed by a perusal of the literature.

As well as involving an extensive frequency profile of ϵ^*, discussions of the dielectric properties of a polymer in the solid state even more commonly rely on the temperature (T) dependence of ϵ^*, and any thorough investigation requires measurements as a function of f and T. The temperature dependence provides information on the activation enthalpy ($\Delta H^{\ddagger}$) and entropy ($\Delta S^{\ddagger}$) for processes, or for some glass to rubber transitions, information on free volume changes. Temperature is complementary to frequency in the sence that when an appropriate scaling law is applied (based on either a free volume theory, i.e., such as that of Williams et al.[110] or a rate process theory such as the Arrhenius equation) the f or T dependence of ϵ^* can be interrelated. This time-temperature superposition principle is of great benefit in relaxation studies[12]. Individual processes, when resolved are conventionally labelled $\alpha, \beta, \gamma \ldots$ in order of increasing frequencies for maximum loss at constant temperature (f_m), or decreasing temperatures for maximum loss at constant frequency (T_m).

The coalescence of some of these processes at certain temperatures is always a possibility since the rate processes generally have differing temperature coefficients. In the polymeric solid state where the coupling of modes is common because of intra- and inter-chain interactions, the equivalence in rate of two different motions at a common temperature frequently leads to a continuance of a coupled process at higher temperatures. One common such occurrence is a combination of the α- and β-process into an unresolved $\alpha\beta$ process[111].

The study of a large number of polymers in the solid state has shown that often similar mechanisms are basic to α-, β-, γ- . . . processes. α-Processes are very frequently associated with the glass to rubber transition, β-processes are often due to local backbone modes of more limited extent than those involved in the α-process, and high frequency/low temperature relaxations (γ etc.) are the result of dipole movement in side chains when in an amorphous or crystalline phase. Although such assignments can be seen in the relaxations of many polymers, they are not universal and should not be taken as a basic concept. In fact the unambiguous assignment of any group of observed relaxations to molecular motions in the solid state can be difficult. Observations of many kinds certainly help in this direction. The use of alternative relaxation techniques[12, 109, 112] such as mechanical, ultrasonic, n.m.r. (both ^1H and ^{13}C) and others frequently help by locating under which perturbations the relaxations are active. For example, a relaxation common in ^1H n.m.r. and permittivity implies the process involves both charge and proton motion. Dielectric methods have the advantage over others in their easy access to an extensive frequency range. Thermal treatment such as heating and quenching to reduce crystallinity, or annealing to increase it, may alter relaxation behaviour and thereby indicate involvement of the crystalline phase[111]. Studying a series of chemically similar polymers may also be informative in assigning similar relaxations to common chemical features[111, 113]. The influence of pressure has been used as a method of indicating that free volume is involved in some relaxations giving direct evidence on main chain motion[111, 114].

Loss peaks obtained for solid polymers are generally much broader than a Debye process, and indeed broader than is the case for many relaxations in solution. In many cases they are grossly unsymmetrical and even show structural features. For broad and even unsymmetrical loss peaks their shapes are often analysed in terms of a standard distribution function as described on p. 106. It should be stressed, however, that anomalous peak shapes can result from an artifact in data analysis when relaxations occur at low frequencies (10^{-5} to 10^{-1} Hz). The conventional technique for this frequency range is to measure the decay current (or less commonly, the charging current) of a discharging capacitor with polymer dielectric, as a function of time. For an applied step voltage V_0 and vacuum capacitance C_0, $\epsilon^*(\omega)$ at angular frequency $\omega = 2\pi f$ is given by

$$\epsilon^*(\omega) = \epsilon^*(\infty) + \frac{1}{C_0 V_0} \cdot \int_0^\infty I(t) e^{-i\omega t} \, dt \qquad (6)$$

where $I(t)$ is the discharge current. The Fourier transformation of Eq. (6) can either be done numerically[115, 116] or analytically by specifying the functionality of $I(t)$. Employing the latter by using either the Hamon approximation[117] or the function of Williams et al.[78] prejudges the shape of the loss peak and should therefore be restricted to locating the frequency of maximum loss. Numerical Fourier transformation is preferable and essential, if accurate data on loss maxima, ϵ' or loss peak shape are required[115].

Since the relaxation behaviour of polymers in the solid state depends on molecular motion which is governed by the environment of the chain, studies of dielectric relaxation is a technique sensitive to alterations of environment. Mention has already been made of the effects of temperature, pressure, thermal history and changes of chemical structure on relaxation behaviour. However, this is not an exclusive list of factors affecting relaxation behaviour. A number of studies of orientated polymers have been reported. Such materials become, as expected, anisotropic in permittivity. The drawing of poly(vinyl chloride) or poly(ϵ-caprolactone) causes shifts in loss position and reduces $\tan \delta (= \epsilon''/\epsilon')$[118]. Similar effects occur in poly(ethylene terephthalate)[119]. Yemni and Boyd[120] report that the loss peak intensity of nylon-6,10 is much reduced in both the parallel and perpendicular directions after drawing at 160 °C. Since the relaxations they studied are generally believed to originate in the amorphous phase, these authors deduce that considerable chain alignment occurs in extrusion, even at temperatures some 100 °C above the glass-rubber transition temperature.

The influence of additives to polymers, particularly plasticizers, naturally reflects their relaxation behaviour. There have been a number of dielectric studies investigating the influence of plasticizers[12, 121] most extensively with poly(vinyl chloride). These have shown not only the changes in the motion of the polymer chain but frequently motions due to the plasticizer molecule, if polar. The technique is then related to the matrix isolation method[122] in which the polymer is the bulk phase. As expected, plasticizers have the effect of shifting the glass-rubber process to higher frequencies or lower temperatures[123, 124]. When polar plasticizers are used further information can often be obtained. For example, in the plasticization of poly(vinyl acetate) with benzyl benzoate, low concentrations of plasticizers shift the glass-rubber transition and alter its intensity but no additional peaks are observed[125]. Increased concentrations of plasticizer results in the appearance of an extra peak which finally becomes so dominant at high plasticizer concentration that the original glass-rubber process disappears. These observations were explained by a coupled motion between segments and plasticizer molecule at low concentrations (the mechanism of plasticization) followed by a separate rotational relaxation of free plasticizer molecules at higher concentration. A similar mechanism was proposed by Hains and

Williams[126] who investigated the plasticization of polystyrene. When this dielectrically rather inert polymer is added to dibutyl phthalate, the single relaxation process of the pure ester is initially broadened, but at polystyrene concentrations above 40%, two broad relaxations develop. Analysis suggests that the low temperature process involves ester molecules which are relatively free to relax their dipoles in whole or in part, whilst the high temperature relaxation involves the cooperative motion of polymer chain and ester dipoles. This latter process thus being responsible for plasticization.

Extraneous molecules in solid phase polymer systems are not limited to plasticizer molecules or even exclusive to substances deliberately added. Impurities when present often affect the dielectric behaviour of polymers and water in particular often has very significant effects on the dielectric spectrum. Poly(methyl methacrylate)[123, 127], poly(oxymethylene)[128], and nylons[129] to mention a few are influenced by moisture in this way. The influence of moisture on dielectric relaxation can be the result of interfacial polarization as well as dipolar mechanism. Further, this complication is not restricted to additives such as water but may occur whenever a combination of phase boundary and bulk or surface conductivity to or over the boundary can take place. The proof that a relaxation is the result of interfacial polarization is not easy to establish, but there is evidence that some of the relaxations in nylons[130] and poly(urethanes)[131] are of this type. As expected, conductive fillers will introduce interfacial polarization and this effect has been well documented, especially in carbon filled rubbers[132]. Indeed, as we shall discuss later, electronic conductance when localized by interfacial boundaries does result in a form of interfacial polarization. Here, because of its large magnitude the phenomenon has been termed hyperelectronic polarization.

4 Thermally Stimulated Discharge

The previous section reviewed the advances of what may be termed the classical technique of dielectric spectroscopy, which has, as its modes of operation, measurements either at constant temperature or at constant frequency. Recently a technique which does not operate with these restrictions has been developed, particularly by van Turnhout[133]. This technique is variously termed thermally stimulated discharge (TSD) or thermally stimulated current (TSC). In its simplest form a material in the form of a disc with electrodes in intimate contact on opposite faces is heated to some 'forming' temperature T_f. At this stage the sample is polarized by applying a strong electrostatic field (≈ 30 kV/cm) for an extended period. All polarization processes which are active at, or below the temperature T_f and within the time the field has been on, will contribute to an overall polarization: the material within the field is polarized. If the field were now removed at the temperature T_f, the polarization could be discharged as in the step response technique of the previous section. However, in TSD the temperature is reduced to some value $T_s < T_f$ with the field still on. Any polarization process which is non-active at T_s but active between T_s and T_f has thereby been frozen-in at $T \leqslant T_s$, even when the field is removed as in the TSD technique. The material so produced with its residual polarization in the absence of the field, is termed an electret. In TSD the polarized sample at T_s is first short-circuited and then connected through an electrometer using the same electrodes as were used for polarizing. Heating at a constant rate, $\sigma_d = dT/dt$ then results in depolarization by the various molecular processes relaxing, and that depolarization, by causing a changing induced field, stimulates a discharging current which is recorded as a function of temperature.

The magnitude of the current density $j(t)$ for a shorted electret depolarizing by a first order process, and with an instantaneous polarization $P(t)$ is given by

$$j(t) = \tau(T) \frac{dP(t)}{dt} = P(t) \tag{7}$$

where $\tau(T)$ is the relaxation time at T. The value of $P(t)$ is determined by the initial charging process under the field E as governed by:

$$\tau(T) \frac{dP(t)}{dt} + P(t) = \varepsilon_0 \Delta\epsilon'(T)E \tag{8}$$

for an increment of relative permittivity $\Delta\epsilon'(T)$. Equation (7) also shows that there will be a peak in the form of a current maximum in the current temperature or current-time profile (inter-related by σ_d). By differentiating Eq. (7) such a maximum occurs when

$$-\frac{d\tau(T)}{dt} = \sigma_d \frac{d\tau(T)}{dT} = 1 \tag{9}$$

The shape and intensity of the TSD peak or peaks (recorded in the current-temperature plane) depends upon the electrical and thermal history as it effects $P(t)$ or the related $P(T)$. For a charge-discharge cycle as shown in Fig. 10a followed by the experiment starting at t_d the resulting TSD curve will show one or more peaks corresponding to various polarizing mechanisms as illustrated in Fig. 10b. In most practical situations the individual peaks will be broader than that due to a single first-order relaxing process. To introduce the analysis of peaks shape we start with a first-order process whose polarization $P(t \geq t_d)$ is given from Eq. (7) by

$$P(t \geq t_d) = P(t_d) \exp - \int_{t_d}^{t} \frac{dt}{\tau(T)} = P(t_d) \exp \left[-\frac{1}{\sigma_d} \int_{T_d}^{T} \frac{dT}{\tau(T)} \right] \tag{10}$$

To find $P(t_d)$ Eq. (8) must be integrated from $t = 0$ (when it is presumed there is no polarization) to t_s, to give $P(t_s)$, followed by integration of Eq. (7) from t_s to t_d. Since $\Delta\epsilon'(T)$ does not change rapidly with T($\Delta\epsilon' \propto 1/T$ for dipolar relaxations) in comparison to the exponential dependence of $\tau(T)$ on T, $\Delta\epsilon'$ is taken as effectively T invariant in the experimental temperature range. Integration then gives

$$P(t_d) = \varepsilon_0 \Delta\epsilon'E \left\{ 1 - \exp \left[-\frac{t_f}{\tau(T_f)} \right] - \frac{1}{\sigma_s} \int_{T_f}^{T_s} \frac{dT}{\tau(T)} \right\} \exp \left[-\frac{(t_d - t_s)}{\tau(T_s)} \right] \tag{11}$$

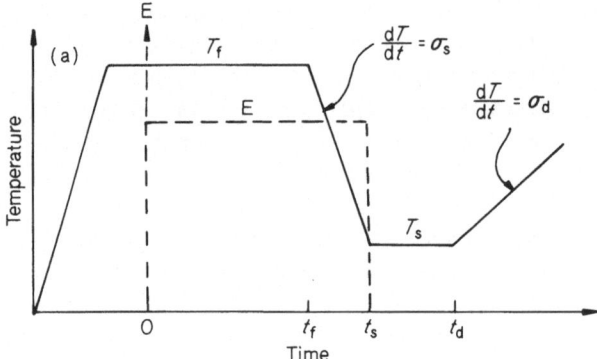

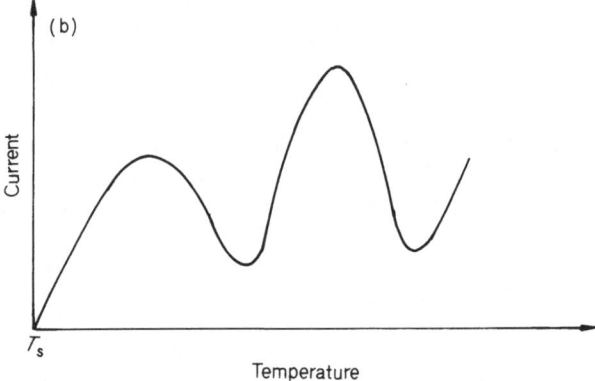

Fig. 10. a The charge-discharge cycle used in a one-step thermally stimulated discharge (TSD) experiment. T_f, forming temperature; T_s, starting temperature; $0-t_f$, forming time; t_f-t_s, shorting time; t_s-t_d, storage time; t_d, start time for discharge. Dashed curve gives the field (E) – time dependence. **b** Illustration of a TSD curve showing two processes

Equations (10) and (11) show that TSD spectra reflect the temperature dependence of the relaxation time $\tau(T)$, or, if the processes can be analytically described by a distribution of relaxation times, their temperature dependencies. The two common forms of $\tau(T)$ are:

(i) an activated process with an Arrhenius dependence $\tau(T) = \tau_0 \exp\left[U^{\neq}/(kT)\right]$ having an activation energy $U^{\neq}$ or,

(ii) a free-volume controlled process obeying the Williams, Landel and Ferry[110] behaviour: $\tau(T) = \tau_0 \exp\{-C_1(T-T_g)(C_2 + T-T_g)^{-1}\}$ where $T_g < T$ and C_1, C_2 are constants.

In these expressions τ_0 refers to the natural frequency as $T \to \infty$ (activated process) or as $T = T_g$ (free volume control). Above T_g the latter dependence degenerates into Arrhenius form in terms of a scaled temperature[133] $T' = T - T_g + C_2$, and even below T_g an effective temperature can be defined to ensure Arrhenius form[134]. Thus for a single relaxation the TSD peak can be fully analysed to provide an activation energy or shift temperature.

For a distribution of relaxation times (or the equivalent non-first-order behaviour) of pre-selected type, the analysis, though more complex is also possible. The reader is referred to van Turnhout[133] for a detailed discussion of such analysis and for the charging-discharging conditions required for sensible analysis to be made. Several distinct cases arise. Firstly, a distribution of processes can result as either a continuum of relaxation frequencies or as a non-first-order rate at constant activation energy, or also involve a distribution of activation energies. A process limited to a constant activation energy is termed thermorheologically simple, and provided also that a specific form of distribution is used to fit the TSD data, the activation energy can be calculated from the half-width of the TSD peak. More complex 'least-square' fitting methods enable the form of distribution and activation energy to be estimated without a specified distribution of relaxation times. When a distribution of activation energies is also involved, the analysis is of even greater complexity and can, in the case of a single direct TSD experiment, only be done for an assumed distribution of relaxation frequencies. However, the experimental possibilities are fortunately not restricted to such a single step TSD experiment in which charging to saturated polarization under the field is followed by a thermally induced total discharge. Various fractional charging-discharging procedures have been described[135, 136] which provide by decomposing a TSD peak, information on the distribution of activation energies and frequencies. Of particular interest is one recently described by Vanderschueren[136] in which the field is applied as a discrete sequence of steps during a slow cooling process from T_f to T_s. When the field is off, the sample is short-circuited. The subsequent curve will then show partial peaks corresponding to the polarization process which occurred over each charging step. Figure 11 shows the decomposition of the β-peak of poly(ethyl methacrylate) achieved by Vanderschueren[136]. Individual

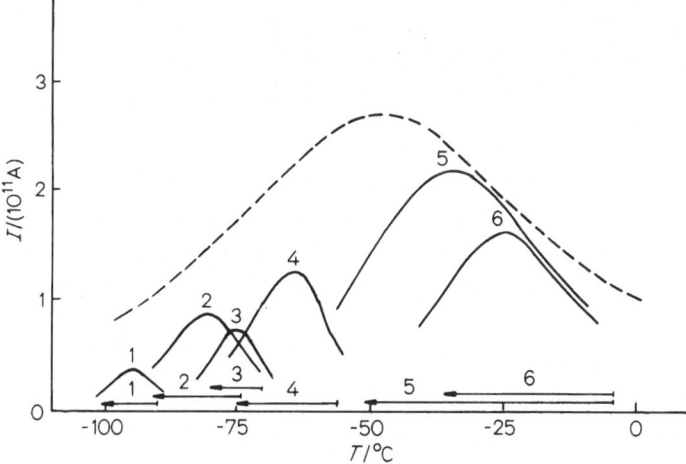

Fig. 11. The multi-step, fractional charging process used by Vanderschueren[136] in decomposing the β-process in poly(ethyl methacrylate). The partial peaks are indicated in terms of the temperature ranges for the polarizing steps (←—). Dashed curve corresponds to a single step TSD on the polymer (conditions for saturated polarization: T_f = 40 °C; T_s = −196 °C, E = 10 kV/cm, t_f = 1800 s)

analysis of each resolved peak provides an activation energy whose temperature dependence is thereby resolved. This method has been used for the analysis of the β-peaks of nylon-6,6, poly(methyl methacrylate), poly(ethyl methacrylate), poly-(tert-butyl methacrylate), poly(vinyl chloride), poly(vinyl acetate), poly(vinyl alcohol), poly(tert-butyl acrylate) and poly(phenyl methacrylate)[136].

As in dielectric spectroscopy, both dipole orientation and interfacial polarization can cause TSD peaks. In the latter context the technique (then frequently referred to as thermally stimulated currents or thermally stimulated conductance, TSC) has diagnostic value in the study of semi- and photo-conductors, a topic further discussed below. However, it should be mentioned at this stage that an absence of good electrical contact between sample and electrodes can result in charge injection into the sample by, for example, electrical discharge over the air gap. On occasion, charge injection is deliberately undertaken to produce hetero-electrets with a net excess charge[133]. With semi-conductors such charge injection can also occur with blocking but contacting electrodes.

In the opinion of the reviewer, the technique of TSD has its greatest merit in being able to rapidly provide information on the number, and to some extent, origins, of low frequency dielectrically active relaxing processes in a polymer. Although analysis of peak shape can provide information on the distribution of relaxation times and activation energies, this generally requires more experimentation and computation. Thus for such information conventional dielectric techniques may well prove superior.

5 Piezo- and Pyro-Electric Polymers

Polymer electrets generated as described for TSD experiments are pyro-electric materials since their polarization changes on heating. They are also to a greater or lesser extent, piezo-electric since the polarization changes under strain either because of a change of dimension (Poisson's ratio) or because of electrostriction (strain dependence of permittivity). The anisotropic charge distribution in such electrets is commonly rather small and of a transient character after a heating cycle. This is because the charge asymmetry has been induced by an electrostatic field capable of only slightly perturbing the charge distribution. For certain materials an intrinsic polarization can be induced which is large, and which has much greater permanence after heating, provided some upper limiting temperature is not exceeded. These intrinsic piezo- and pyro-electric materials frequently obtain their "permanent" anisotropic polarization by some structural re-arrangement involving either crystal packing or dipole alignment of macro-dipoles. Poly(α-amino acids)[137-139], cellulose acetate[139] and certain bio-polymers[137] are examples involving the latter mechanism. Poly(vinylidene fluoride)(PVDF) has become the best known polymer belonging to the former category, following the original observations of Kawai[140] and Bergmann et al.[141]. However, other fluorine containing polymers and copolymers, such as poly(vinyl fluoride)[142], copoly(vinyl fluoride-vinylidene fluoride)[143], and copoly-

(tetrafluoroethylene-vinylidene fluoride)[143, 144], owe their piezo- and pyro-electric properties to their morphology.

Although a study of the piezo- and pyro-electricity of a polymer can provide valuable information bearing on polymer morphology and properties, this review will concentrate on the applications of these materials with particular reference to PVDF. For a fuller discussion of the basic theory the reader is directed to the excellent reviews by Hayakawa and Wada[145].

PVDF film electrets are prepared in two stages by first stretching the film and then poling. The former process, which can be done uni-axially or bi-axially at a temperature in the range 60 to 100 °C, interconverts the crystal α-form of the crystallites to the β-form (Fig. 12), and probably increases the extent of crystallinity of this already highly crystalline material (40–60%). Subsequently electrodes are evaporated onto the film and the material poled under a field of 100–3,000 kV/cm for ≈ 20 min at 80–90 °C, followed by cooling to ambient in the field. The resulting film is highly polarized and although some of that polarization is lost in the first heating cycle, considerable polarization is well retained at temperatures up to 70 °C, and the decay is slow at temperatures up to 130 °C. The polarized PVDF film does not have a large external electric field associated with it as might be expected, because the forming conditions are such that surface charges are injected, counteracting the bulk polarization. However, a change of temperature, or the application of stress induces a change in bulk polarization and consequently a field change. It is this property which provides the mechanism for PVDF film devices.

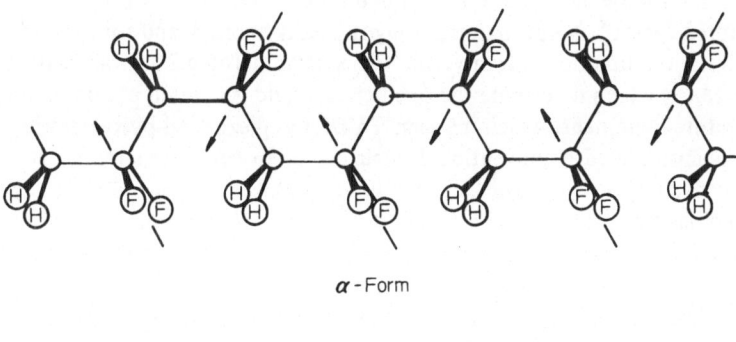

α - Form

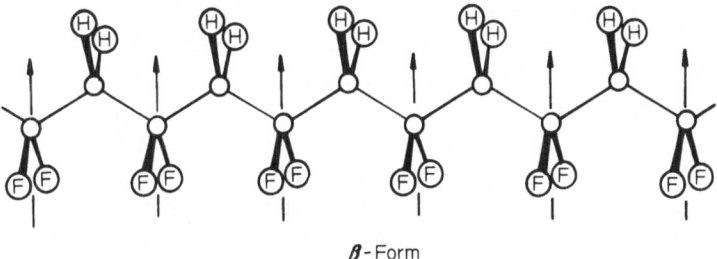

β - Form

Fig. 12. The α- and β-chain forms of PVDF

The magnitude of these effects are determined by the piezo-electric strain con-
stand $d_{j,k}$ or pyro-electric constant $p_{j,k}$ defined in expressions (12) and (13) respectively.

$$d_{j,k} = \left[\frac{\partial^2 G(X,E,T)}{\partial E_j \partial X_k}\right]_T = \left(\frac{\partial D_j}{\partial X_k}\right)_{T,E} = \left(\frac{\partial S_k}{\partial E_j}\right)_{X,T} \tag{12}$$

$$p_j = \left[\frac{\partial^2 G(X,E,T)}{\partial E_j \partial T}\right]_X = \left(\frac{\partial D_j}{\partial T}\right)_{X,E} \tag{13}$$

Thus these coefficients arise from the cross-variation of the free energy (G) of
an electret with E and stress (X) or with E and T. Because of the anisotropic situation
inevitable in an electret, d and p are tensor quantities defined by the axes (j = x, y, z;
k = xx, xy, xz, yy, yz, zz). Since a change $\partial G/\partial E_j$ manifests itself as an electric displacement ($D_j = E_j + P_j$) and $\partial G/\partial X_k$ as a strain (S_k), the subsequent equalities
result. It can be seen from definitions (12) and (13) that the geometrical situation
is complex. There are other complications in a variation of definition which depend
on such differences as to whether the free or clamped piezo-electric constants are
being used, or whether thermal expansion is or is not included in p. For the comparison purpose pertinent to this review it is not necessary to be deeply involved in
these intricacies. For a more detailed discussion the reader is referred to Broadhurst
et al. [146]. The electromechanical coupling constant is small and since polymer films
are involved, the useful constants are preordained as j normal to the film and k in its
plane. Table 2 shows some data of d and p quoted by Broadhurst et al. For quartz
the value is quoted for a 0°, X cut material, longitudinally excited, and for barium
titanate the excitation direction is transverse. Also shown in Table 2 are values of
$d/(\varepsilon_0\epsilon')$ and $p/(\varepsilon_0\epsilon')$ which determine the open circuit field generated per unit stress
or unit temperature increment. As can be seen, PVDF has piezo- and pyro-electric
properties commensurate with conventional single crystal or ceramic materials. It
also has certain advantages: it is lighter, can easily be produced as a thin film of large
surface area, and has a high internal resistance. It is also cheap to produce. Admitted-
ly its mechanical stability does not match single crystals or ceramics, but its very
flexibility can, for some applications be an advantage.

The mechanisms which render PVDF and other fluorine substituted members
of the polyethylene class such outstanding piezo- and pyro-electrics is related to
their structure and morphology. PVDF is a highly crystalline material having three
different possible crystal structures (α, β and γ). Of these, the α-phase is formed
from the melt at $\leqslant 150\,°C$ whilst the γ-phase is formed at $\approx 170\,°C$. Extension of
the α-phase produces the β-form and it is this form which is the most piezo- and
pyro-electric. Figure 12 shows a representation of the chain configurations in the
α- and β-crystal forms and it can be seen that the extended chain structure of the
β-form results in a large summative dipole moment, whilst the α-form because of
symmetry, has a much reduced summative dipole moment. There is evidence that
the piezo- and pyro-electricity of PVDF together with its large polarization under
a field are dependent on the number of crystallites in the β-phase, and the ease with

Table 2. Piezo- and pyro-electric characteristics of some polymeric and non-polymeric materials [146]

Material	$d/(10^{-12} \, mV^{-1})$	$(\varepsilon_0\varepsilon')^{-1}d/(10^{-3} \, mV \cdot N^{-1})$	$p/(10^{-9} \, C \cdot cm^{-2} \cdot K^{-1})$	$(\varepsilon_0\varepsilon')^{-1}p/(10^4 \, V \cdot m^{-1} \cdot K^{-1})$	Ref.
Quartz	2	50	–	–	147)
Barium titanate	190	12	20	1.3	147, 148)
Triglycine sulphate	–	–	30	100	148)
Poly(vinylidene fluoride)	30	350	3	34	147, 148)
Poly(vinyl fluoride)	1	–	0.26	–	146)
Poly(vinyl chloride)	1.5	63	0.3	11	146)

which these can align their polarization. Also important in the mechanism of polarization is the ability of residues to be incorporated into the crystallites from amorphous regions, and to interconvert their orientations at grain-boundaries. The bulk
of the evidence supports the view that the large polarization upon which piezo- and
pyro-electricity depend, are, in the case of PVDF the result of dipole orientation.
Some charge injection does appear to take place in the poling of PVDF film[149], but
it is now generally believed that bulk polarization by an interfacial mechanism is
not the major process responsible for piezo- or pyro-electricity.

It is also significant that other partially fluorinated polyethylenes are in general
very piezo- and pyro-electric. A combination of factors are thought responsible. The
high electronegativity of fluorine generates large residue dipoles (unless the substitution pattern is such that symmetry negates this), and its small bulk makes conformation rearrangements relatively easy. Polarization by residue dipole reorientation
becomes relatively easy, and its magnitude and permanence is greatest when crystal
forms are present which stabilize a favourable structure, such as the β-form of PVDF.

The applications of PVDF and other fluorinated analogues as piezo- and pyro-
electric materials is a growth area. Broadband ultrasonic transducers have been constructed from PVDF which respond to a very wide frequency band[140, 150, 151]. They
can also produce wide-aperture sonic beams as shown by Bui et al.[150] who demonstrated transmission between rectangular (6.4 x 19.2 mm) PVDF sheets bonded to
copper acting as transducer and receiver. The material has been used to construct
lightweight microphones[150, 152, 153] and in loud speaker systems used for audible
warning devices[153]. The modulation of light by ultrasonic waves using PVDF films
has been described[154] as has its application in contactless pressure switches where
several volts can readily be produced by finger pressure[153, 155].

PVDF is at least equally versatile as a pyro-electric material in infra-red detectors.
Its large potential aperture, or its ability to provide arrays of sensors by metallizing
the film through masks, provide great versatility. Infra-red absorption is achieved
either by the polymer, or by the front electrode, or by blacking the material to give
the maximum absorption. The electrical time constant for the response is slow making it possible to use it in a quasi-static fashion. However, by thermal loading using
a metal backing a faster response can be achieved, albeit with some loss of sensitivity.
As well as its use for a direct detector[156] of large aperture it has been used as a vidi-
con target[153, 157], in laser beam profiling[153], for reflectivity measurements[158], and
in electrostatic copying machines[159].

6 Conducting Polymers

The subject of conductivity in polymers has been extensively studied and reviewed[160−162]. It is a field in which the complexities of solid state physics and synthetic
chemistry intermingle, and for this reason presents inter-disciplinary problems in
their most accute form. Many polymers have been synthesized with a view to producing good conductors or semi-conductors which retain the desirable polymeric
attributes of moldability, flexibility and toughness. Basically, these properties tend

in some degree to be incompatible. The result is that many macromolecules whose structure on paper may give hope to the chemist that undertaking a synthesis would produce a good conductor, give, in the event, a material of poor conductance, or mechanical properties, or both. The extent of any disappointment is determined by the expectations, and frequently by the somewhat arbitrary scale of conductance which are said to define an insulator, semi-conductor or conductor. In the opinion of the reviewer the conducting properties of a polymer can firstly be of basic scientific interest as a solid state property, and secondly, may be of practical use in the materials technological application. For certain uses requiring a conducting polymer the high conductance of a metal may not be essential. To cite two examples. The moderately conducting polymer system of quarternized poly(2-vinylpyridine) doped with tetracyanoquinodimethane discussed on p. 139 has been patented for the use in electrical circuitry[163]. The other example concerns high fidelity coaxial cables. These are provided with a semi-conducting sleeve between braiding and core-insulation so as to provide improved screening and eliminate electrical noise due to any mechanical movement of the cable. Although the sleeve is commonly made from a carbon loaded plastic, it has been suggested that an in-situ prepared semi-conducting surface be used, which is based on condensing nylon-6 with formaldehyde and 4-vinyl-pyridine followed by charge-transfer complexing with boron trifluoride[164]. Further uses or potential uses of conducting polymers will be given later in this section, and that of photo-conducting polymers deferred to the next section. However, excluded from these discussions are composites which are conducting because of a loading of carbon or other traditional conductor. The technology of these very useful materials have been reviewed[165].

The conductance of a material is measured in terms of its conductivity σ in ohm$^{-1} \cdot$ cm^{-1}, and it is well known that σ depends upon temperature. It also depends upon pressure, and frequently in the case of polymers significantly so, because packing, particularly in powdery materials, effects the ease of charge migration. Further σ can be, or appear to be a function of the electrical field strength. Implicit if this is so, is that either the sample or the interface between electrodes and sample are non-ohmic. Both the effect of pressure and field strength can give rise to complexities in measurement. It is good practice to make measurements as a function of pressure, particularly for samples which are in the form of compressed powders. The ohmic behaviour or (otherwise) of the system should also be checked. Except for the poorest of conductors and those in which charge is carried ionically, most semi-conducting polymers are ohmic at low field strengths when good contacting noble metal electrodes are used. High fields are more prone to produce non-ohmic behaviour and this behaviour can be used diagnostically, particularly with photo-conductors as described below.

The variation of σ with T is of great significance and can be made on the basis of differentiating between metallic conductance and semi-conductivity. Metallic conductance is characterized by a decrease in σ with T whilst for semi-conductors the reverse is the case; σ is often exponentially dependent upon T showing an activation mechanism for conductance with an activation energy $U^{\neq}$.

Comparisons in conductivity are made in terms of σ defined at a reference temperature (frequently near ambient) and the spread of values shown by materials is

Table 3. The electrical conductivities of some materials

Substance	$\sigma/(ohm^{-1} \cdot cm^{-1})$	T/K[a]	sign of $d\sigma/dT$[b]
Quartz	$10^{-17}-10^{-20}$	a	+
Sulphur	$\approx 10^{-16}$	a	+
Polystyrene	3×10^{-14}	a	+
Nylon-6,6	$\approx 10^{-14}$	a	+
Poly(N-vinylcarbazole)	$\approx 10^{-14}$	a	+
Selenium	$(2-4) \times 10^{-6}$	273	+/−
Germanium	11.2	273	+/−
Silicon	12−50	273	+/−
Graphite	10^3-10^4	273	+
Nichrome	9.1×10^3	293	−
Lead	5.2×10^4	273	−
	$\approx \infty$	4	
Copper	5.9×10^5	273	−

[a] a: Insensitive to small T changes, ambient quoted.
[b] +/−: Exhibits differing sign over various T ranges.

very large. The conductivities of semi-conductors are normally taken to be in the range 10^{-10} to 10^2 ohm$^{-1} \cdot$ cm^{-1}, materials are considered insulating at $\sigma < 10^{-10}$ ohm$^{-1} \cdot$ cm^{-1} and become metallic with $\sigma > 10^2$ ohm$^{-1} \cdot$ cm^{-1}. Table 3 quotes representative values of σ for a number of materials.

For electrical conductance to occur it is necessary for there to be charged species acting as carriers and for these species to have mobility through the bulk of the material. Although ions can support the passage of current in the solid state, such conductance is seldom ohmic or time invariant because of the polarization of the electrodes and concentration changes of carriers in the material. However, of greater interest and importance are conductors, including polymers, in which the carriers are electrons or electron holes, since these tend with appropriate electrodes, to provide materials which are ohmic, and most important, not subject to electrolytic change. The conductivity of a material having mobile positive (hole) and negative (electron) carriers of concentration n_+, n_- and mobility μ_+, μ_- is given by

$$\sigma = (n_+\mu_+ + n_-\mu_-)e \tag{14}$$

where e is the electronic charge. Both the number of carriers and their mobility can be temperature and field dependent; the major influence of pressure on σ in polymers is on mobility. Usually the presence of an activating process is taken to indicate that the density of mobile carriers is being increased ($n = n_0{}^{-U^{\neq}/(kT)}$). Mobility is also T dependent, often increasing temperature reducing μ by phonon interference. As developed below these divisions are not so clear-cut when detailed mechanisms are proposed.

Most experimental investigations into the conductance of polymers are limited to measurements of σ and its T and less often, pressure dependence. To measure σ

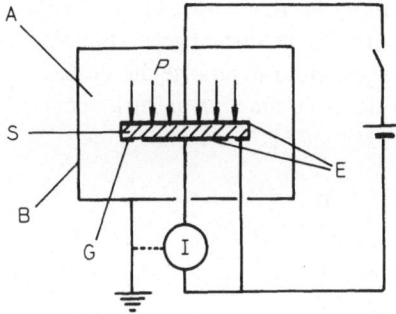

Fig. 13. Schematic diagram showing important aspects of conductivity measurements in polymers. B, screened box enclosing vessel with thermostatting and inert atmosphere A; sample S under pressure P between electrodes E which include guard ring G; ammeter (electrometer) I

equipment of the type schematically illustrated in Fig. 13 is used. Points of importance are: i) to ensure that good electrode contact is made (most effectively by evaporating electrodes onto the surfaces, and ii) that surface conductance is eliminated from measurement by a guard ring. Measurement of sample thickness (required in calculating σ from current data), particularly with films obtained by solvent casting can present problems. Air capacitance measurements at identical spacing or directly by means of a second mechanically coupled air condensor provide an alternative to micrometer measurement. The determination of individual mobilities, or even the sign of the majority carriers is undertaken much less frequently than σ measurements. This is a reflection of the experimental difficulties involved consequent upon the very low mobilities of carriers in polymers. Hall effect studies which measure the potential developed by current displacement of mobile carriers in a magnetic field (Fig. 14 a) are, for most polymeric semi-conductors stretched to the limits of sensitivity because of the low values of Hall mobility, μ^H, although the development of a.c. methods and multiple electrode systems has done much in recent years to improve the technique[166]. Sample shape in the conventional Hall technique also presents problems for most polymeric materials since it is much easier to prepare a film or disc than a rectangular block. The Corbino-disc geometry using an a.c. signal between central and peripheral electrodes (Fig. 14b) can be used[166, 167]. The presence of the magnetic field normal to the disc induces a circular component of current which can be separated from the radial drift current and detected by a coaxial pick-up coil. It is a measure of the Hall mobility. The thermoelectric effect Fig. 14 c and its Seebeck coefficient Q (the voltage developed per unit temperature difference) can be used to estimate mobilities, and particularly, to obtain the sign of the majority carrier. Since carriers travel from the hot side, the polarity of the electrodes indicates the sign of the majority carrier. For photo-conductors, non-uniform optical-excitation produces a gradient of carriers (Dember effect) whose sign and decay can provide information on the nature and mobility of the majority carrier. For the measurement of mobilities in polymers all these effects are insensitive because sample resistances are too high (frequently due to low mobility rather than a lack of carriers). A group of techniques in which low carrier mobility is a virtue rather than a vice are the 'time-of-flight' or 'transit-time' techniques. In principle these techniques measure the transit time for a group of carriers to drift under a field across a known width of semi-conductor. For low mobility, this time scale can be substantial and accurately measured. Various methods exist for the surface generation of either a narrow band of carriers which travel as a packet, or to arrange the sudden and subsequently continuous generation to provide a step-function in carrier con-

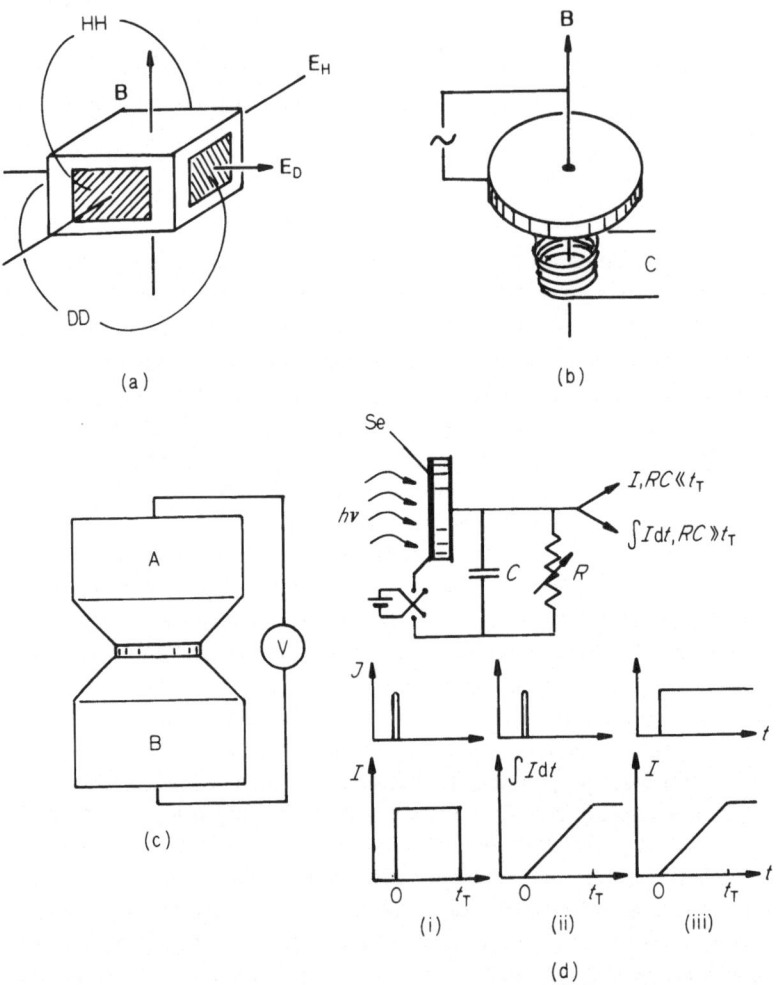

Fig. 14a–d. Methods of finding the charge and mobility of a carrier. **a** Hall effect: E_D; Drift field across electrodes DD on a rectangular parallelepiped sample; B, magnetic induction; E_H, Hall field detected across electrodes HH. Hall mobility μ^H from $E_H = \mu^H B \cdot E_D$. **b** A.c. Hall effect using Corbino-disc with electrodes at centre and circumference; circular Hall current detected by 'pick-up' coil C. **c** Seebeck effect with sample between heat sinks A and B at $T_A > T_B$. The Seebeck coefficient $Q = V/(T_A - T_B)$. **d** Time-of-flight methods. Se, selenium front electrode under irradiation (intensity J) injects carriers into the sample, and their subsequent rate of drift under a field is detected. Depending on C, R and the transit time for charge to traverse the sample (t_T) the current (I) or total charge build up ($\int I \, dt$) can be monitored. Case (i), photon pulse with I detection; case (ii), photon pulse with charge detection; case (iii), photon step with I detection

centration (Fig. 14 d). For instance, photoemission from an appropriate surface electrode subsequent to a light flash or sudden illumination by means of a shutter can be used. The drift of the packet or front of carriers in a field is then monitored electronically. For photoconductors the surface can be charged in the dark by corona discharge and its drift under illumination determined. This is an application of the technique of electrophotography or xerography described in Sect. 7. The sign of carrier can be selected by the nature of the photogenerating electrode or surface charge. Many variants of the technique are known, and in application a number of factors such as carrier life-time and carrier diffusion are important parameters. For a more detailed description the reader is referred to an excellent article by Dolezalek[166] and to the references quoted therein. It is hoped that these developments will result in increasing our knowledge of mobility in polymeric semi- and photo-conductors, since our understanding of such systems certainly requires this type of information.

Electronic conductance in the solid state is for regular crystalline solids most adequately discussed in terms of band conduction. That is the thermal (or photo-physical) promotion of electrons into the conductance band leaving holes in the valence band, with one or both of these species having spatial mobility to migrate in a field. For a disordered solid, which is the case for polymers, such a simple picture is obviously inadequate. If the language of band theory is still preferred, the attenuation of the conductance due to disorder is accounted for by introducing trap states. These are sited energetically between the conductance and valence bands and provide an opportunity for the carrier to be trapped. Regeneration from these traps can take place (the probability depending upon their depth) and the carrier then continues to transport charge. Because trapping and the residence of carriers in traps increases the time required for a carrier to traverse through the material, the mobility is descreased. The number of carriers depends on the energy gap between conductance and valence bands and also between the various trap levels and these bands. The model makes no a priori comment about where traps may be spatially situated (as distinct from their level energetically), although it is probable that irregularities in morphology such as crystal dislocations, crystal-to-amorphous interfaces, and points of irregular chain configuration in amorphous regions, make up such sites. An alternative to the band theory often favoured for polymers is the hopping model for conductance. Here the carrier is circumscribed in its spatial mobility to domains in the structure and can only move to a neighbouring domain by overcoming an energy barrier. The energy barrier to be overcome is determined by the energy levels available to the carrier in its domain and the distance required for the hop to the next domain. Pictorially these models are not very different. In the band model for a solid with disorder, traps, although energetically defined, have spatially limited the extent of the perfect band. For the hopping model the energy levels of the domains are associated with resonance orbitals or bands and the problem of transport lies in progression between domains. In both models, imperfections or irregularities in structure lead directly to difficulty in conductance. For the polymer chemist the focus becomes clear. Disorder and non-regular structure is the anathema of conductance, although highly ordered structures may not be good conductors if there is no provision for thermally promiting electrons. It is with these precepts that we shall review the conductance of selected polymer systems.

6.1 Conjugated Polymers

Conjugated π electron systems have molecular orbitals in which the highest filled levels and lowest unoccupied levels are separated by an energy gap whose separation (Δ) decreases with the extent of conjugation. The possibility of the thermal promotion of electrons as a pre-requisit of electronic conductance appears promising for an extended conjugated system such as a polyalkyne. If in such an alternating single-double bond chain all bonds were of equal length and perfectly conjugated along its entire length, it would certainly be easy to thermally promote electrons for polymers of degree of polymerization > 300. This estimate stems from equating such a polyalkyne $\{CR=CR'\}_n$ to a particle in a box situation in which

$$\Delta = 4.75\,(2\,n + 1)/n^2 \ \mathrm{eV} \tag{15}$$

and $\Delta = kT = 0.026$ eV [4] at $T = 300$ K when n = 368 (Goodings[162])). Such extensive conjugation presents great problems to the polymer chemist. Conventional polymerizations in solution which produce polymer whose organization in the solid state has to be achieved subsequently, is not likely to produce the required regularity. Most striking in this respect is the failure to produce significant semi-conductivity in poly(phenylacetylene) and its substituted analogues. Table 4 reproduces some conductivity data on these materials. They are essentially insulators although considerable conjugation is present as shown by their electronic spectra[168] and electron spin resonance spectra[169]. The situation is not helped by the mechanistic ramifications of the free-radical polymerization of phenylacetylenes: short chains caused by a combination of efficient transfer to monomer plus growing radical stabilization by the conjugation of the growing chain. An informative study by Hartel et al.[172] of the conjugated oligomeric thiophenylene-ethylene(I) shows a strong dependence of

---

4   In SI units: 1 eV $\approx 1{,}602 \times 10^{-19}$ J

Table 4. Selected polyalkynes, $\{CR=CR'\}_n$ and their conductivities σ

R	R'	$\sigma/(\mathrm{ohm}^{-1} \cdot \mathrm{cm}^{-1})$	T/K	Ref.
H	phenyl	$\gtrsim 10^{-18}$	293	168)
H	p-aminophenyl	9×10^{-17}, 3×10^{-16}	343	169)
H	p-nitrophenyl	10^{-15}	343	169)
H	p-formamidophenyl	10^{-14}	343	169)
H	p-methoxyphenyl	$10^{-14} - 10^{-18}$ a	293	168)
H	p-chlorophenyl	10^{-18}	293	168)
H	p-phenoxyphenyl	$> 10^{-18}$	293	168)
Phenyl	phenyl	10^{-14}	RT b	170)
H	nitrile	10^{-17}	293	168)
H	H	10^{-3} c	—	171)

a   Variation with method of polymerization.
b RT: Quoted as room temperature.
c   Stereoregular polymer as a thin film.

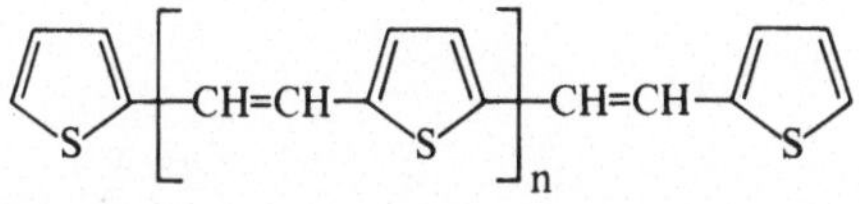

(I)

σ on chain length. They found an increase in log σ from -12.6 (n = 0) to -6.1 (n = 5) followed by a subsequent decrease in log σ to a value ca -7.5 for higher oligomers. The maximum conductivity was shown by the longest oligomer which crystallized (n = 5); for n > 5 the materials were only partially crystalline.

Certainly polymer regularity can be improved by avoiding the mechanistic difficulties in the polymerization of phenylacetylene and its substituted analogues. Shirakawa et al.[171] have synthesised thin films of stereoregular cis- and trans-polyacetylene and these materials in their natural state are much better conductors ($\sigma \approx 10^{-3}$ ohm$^{-1} \cdot$ cm^{-1}). As discussed on p. 117 doping causes a further dramatic increase in σ. Another approach in reducing breaks in conjugation due to chain folding is to lock the structure in the form of ladder polymers. There are a number of

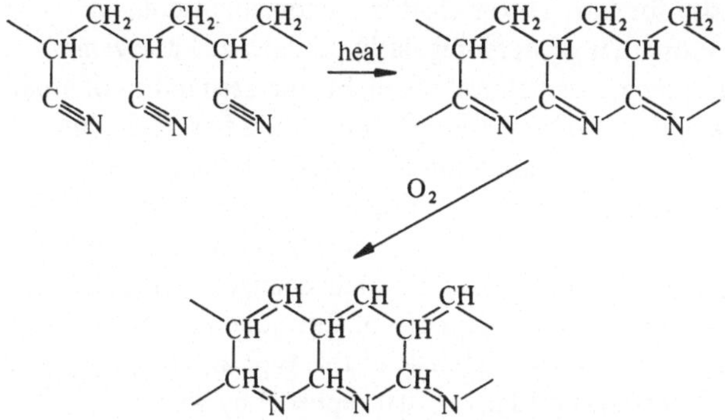

(II)

examples in the literature. Pyrolysis of polymers can cause condensations leading to ladder structures. As illustration, polyacrylonitrile develops structure II ($\sigma_{293} \approx 10^{-9}$ ohm$^{-1} \cdot$ cm^{-1})[173, 174] and the polycondensation product (III) of β-chlorovinyl ketone can attain conductivities[174, 175] of $\approx 10^{2}$ ohm$^{-1} \cdot$ cm^{-1}. Certain

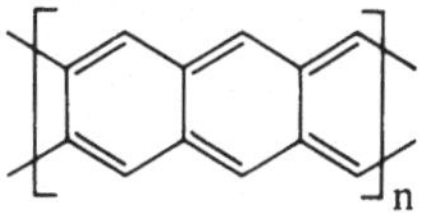

(III)

condensation polymerizations lead to conjugated ladder polymers. Pohl[176, 177] has investigated those obtained by reacting pyromellitic anhydride with multinuclear aromatic compounds to give polyacene quinone radical polymers (as an example structure IV with $\sigma = 3 \times 10^{2}$ ohm$^{-1} \cdot$ cm^{-1}). Such ladder polymers certainly pro-

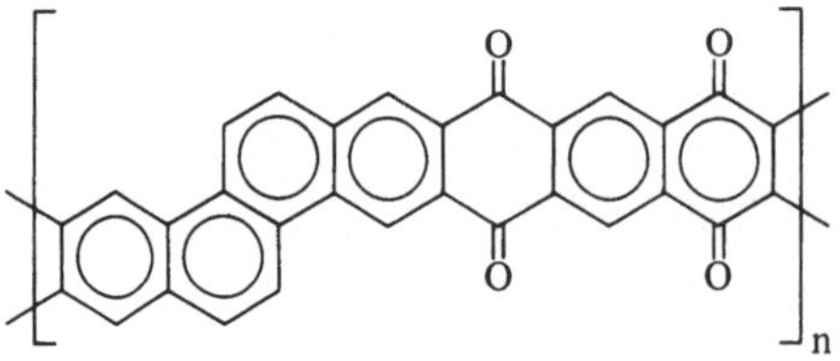

(IV)

duce better conductors than the more flexible conjugated polymers, but they are inflexible, insoluble and as plastics, very intractible. As materials they are part of the way to graphite and approach its conductivity. Present in the bulk materials are potential carriers in the form of radicals or radical ions as evidenced by E.S.R. activity. For the polyacene quinone radical polymers, Pohl and coworkers[176] have shown extreme values of relative permittivity ($\epsilon' \leqslant 300,000$) which are the result of enormous interfacial polarizations across long conjugated domains in which the radical ions are mobile. The phenomena was termed 'hyperelectronic polarizability' by Pohl. These systems show a hopping mechanism of conductance, where in our terminology the domains over which there is a very high mobility and bulk transport are limited by the necessity for carriers to transit from domain to domain.

The superior conductive properties of stereoregularly polymerized acetylene over other conjugated polymers mentioned above, highlights the desirability of order in such a polymer. Solid state polymerizations can provide a route to such highly ordered structures provided that the polymerization proceeds with the minimal introduction of defects. The conditions necessary for such a desirable state of affairs to be possible have recently been reviewed[178]. They involve the criteria that in the polymerization the positions of atoms in the monomer shall change minimally when incorporated into the polymer, and that the symmetry of the monomer shall be such that a unique chain must result. One class of materials of this type are the poly(diacetylenes) whose 1—4 photopolymerization (Fig. 15) first reported by Wegner[179, 180] results in bronze coloured, highly defect free, polymer single crystals. Polymer crystals of considerable size (up to 20 cm long) can be prepared[178]. X-ray investigation[181] has established that the polymer is highly crystalline involving planar con-

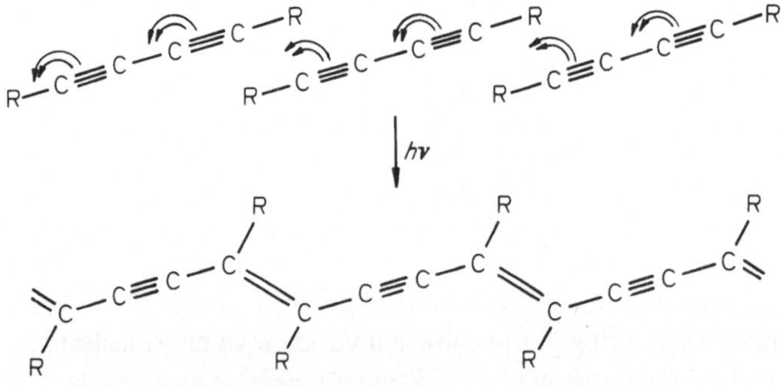

Fig. 15. Poly(diacetylenes) and their photopolymerization

jugated chains with bond alternation and few defects. The crystals are anisotropic in electrical[178], and optical[182] properties. Along the chain direction they also have a very high Young's modulus[183]. The mechanical properties, low concentrations of defects, and molecular weight estimates on homologues in which the side chain is flexible enough to introduce solubility, indicate that the chains extend considerable distances along the major crystal axis[178]. Yet along chain conductivities are very low ($\sigma_{300} \approx 10^{-15} - 10^{-10}$ ohm$^{-1} \cdot$ cm^{-1}) and only some six fold higher than the across chain conductivities[179, 184, 185]. Poly(diacetylenes) are large band gap semi-conductors[179, 184, 186, 187] whose experimental energy gap of ≈ 2.0 eV corresponds closely to that estimated for an infinite chain of alternating bonds[187]. The bond alternation which is a form of Jahn-Teller stabilization in molecular terms, or Peierls distortion in solid state physics terms (to be discussed below) restricts the electron delocalization and separates energetically the conductance from valence bands. Ther-mal promotion of electrons is difficult, although when promoted photochemically the resulting carriers have a high mobility[178, 185] (p. 157). Thus the photoconductive properties of poly(diacetylenes) proves to be of more interest than its semi-conduc-tance.

Another monomer susceptible to solid state polymerization and producing a conjugated chain polymer is di(sulphurnitride), S_2N_2. The resulting poly(sulphur-nitride), $(SN)_n$, is an electron deficient material with an alternating bond structure.

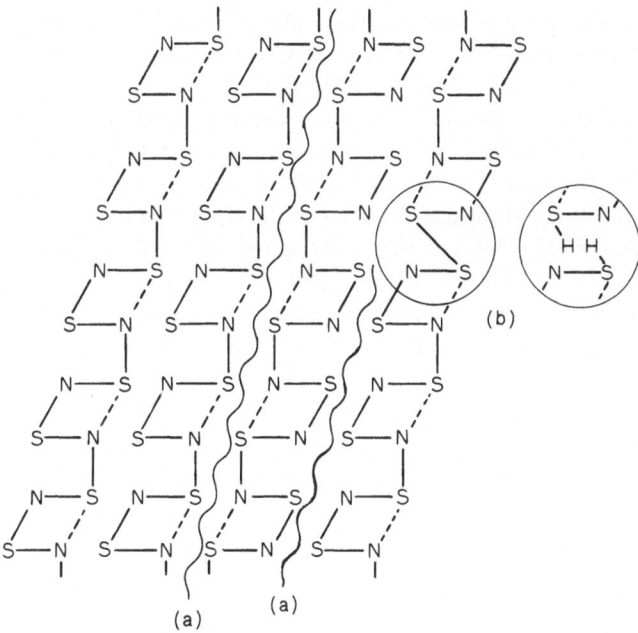

(b)

(a)   (a)

Fig. 16a–b. Schematic diagram of the stacking of $(SN)_n$ resulting from the solid state polymer-ization of the centro-symmetric monomer, S_2N_2. The $(S–N)$ monomer bond which is opened is shown dashed and can lead to defects of type (a) and (b). The formation of $S–H$ (or $N–H$) groups from an incipient type (b) defect in the presence of hydrogen containing impurity is shown in the insert

On the molecular level much more disorder is present than is the case with poly-
(diacetylenes). Undoubtedly the major cause is the non-uniqueness of structure
which can result when a centro-symmetric monomer such as S_2N_2 rearranges bonds
to a chain structure as shown in Fig. 16 at (a). The presence of such mirror image
chains has been found in $(SN)_n$ crystals[188]. The extent of such disorder, which in-
fluences the electrical and other properties of the material must depend upon the
conditions, such as temperature, of the polymerization. Defects of type (b) shown
in Fig. 16 may also be present producing either S-S plus N-N bonds or being terminated with impurities. The latter would be consistent with the observed high level
of hydrogen derived from extraneous impurity[189]. Free radicals are observable
during chain growth but not in the final polymer[190]. In view of the more disordered
state of $(SN)_n$ vis-a-vis the poly(diacetylenes) it might be expected that its conductivity would be poor. However, this is not the case. Poly(sulphurnitride) has a σ_{300}
of 1.1×10^3 ohm$^{-1} \cdot$ cm^{-1} along the chain direction which increases with a reduction
in temperature [191–193]. That is, the material exhibits metallic conduction. In contrast, the across chain conductivity is much lower, increases with temperature, and
in this direction the material behaves as a semi-conductor at 300 K[191–193]. At low
temperatures (<1.4 K) a superconductivity transition occurs, as originally observed
by Greene et al.[194], who also reported that superconductivity can occur in all directions. The transition is of great interest. Its temperature of onset and the supercon-
ductive properties are strongly dependent upon the state of crystal perfection[192, 193, 195],
low defect levels favouring a higher transition temperature. It has been stated[178]
that defect-free $(SN)_n$ should have the highest transition temperature of 1.4 K; the
highest values presently observed have been 0.35 K. At room temperature the conductivity is less than expected from Drude analysis of the optical frequency absorption[192, 196] and this is conceptually consistent with the model of a conducting phase
only slightly limited by defects. The different conductance behaviour between $(SN)_n$
and poly(diacetylenes) stems from the fact that for the former bond length alternation is not so severe, due to the mixed valence character of the chain backbone
which reduces the Peierls distortion. This is an important factor since it shows that
striving for structural perfection with formal conjugation does not necessarily lead
to good conductance. The next group of polymers to be described, those that depend
on charge transfer complexing also are strongly influenced by such factors.

6.2 Charge Transfer and Radical Ion Polymers

It has been known for over a decade that charge transfer (CT) complexes have conductivities much larger than either the donor or acceptor partner on its own, or indeed than would be expected for organic molecules which form the basis of many
donors and acceptors. Complexes of this type have become of increasing interest as
more and more significant levels of conductivity have been achieved, particularly
with ion radical salts. Formally these can be considered as CT complexes in which
total electron donation has occurred. These materials, now classified as 'organic
metals' have been extensively studied and reviewed. Although not strictly within
the purvue of the present review, an understanding of the conductance process in

organic metals is of importance in correlating the behaviour of polymeric analogues. I am particularly indebted to the review by Perlstein[197] which provides an instructive discussion of the mechanism of conductance in organic metals, and on which some of what follows is based.

To enter the discussion of conductance in organic metals it is instructive to investigate the molecular orbital (MO) state of a linear array of iodine atoms and the effect of introducing electrons into such an array. This also has practical significance because iodine and other halogens are common acceptor molecules in organic metals and polymeric charge transfer semi-conductors. Figure 17 a shows the MO energy diagram for linear arrays of two, three, four, and an infinite chain of iodine atoms, equally spaced. The energy gap between the lowest bonding orbital and highest anti-bonding orbital remains essentially constant (four times the one-electron exchange integral), and in the infinite array there are an infinite number of filled orbitals up to the non-bonding band and a similar number of empty orbitals above that level. Thus I_n has a filled valence band adjacent to an unfilled conductance band and would thus be metallic. In deriving this picture, one 'p' electron per iodine atom is used in bonding. The situation depicted in Fig. 17 a is however of higher energy

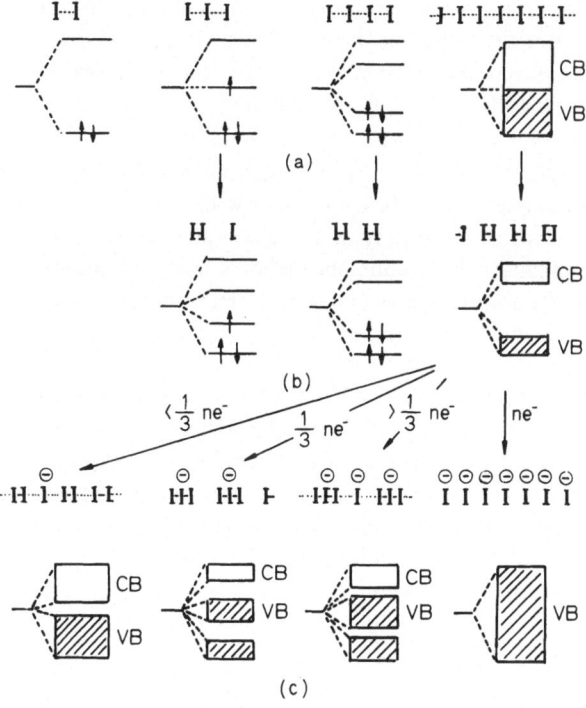

Fig. 17a–c. The energy levels in a linear array of iodine atoms I_n. a Situation for centres equally spaced. b Peierls distortion (Jahn-Teller effect) leading to a lower energy via alternation in bond lengths. c The effect of introducing electrons into system (b). VB, Valence band and CB, conductance band. (For the I^- case there is an upper CB not shown)

(except for I_2), than the situation in Fig. 17 b in which an alternation of bond length occurs. The ranges of energies for the filled and unfilled levels are reduced, finally leading in an infinite chain to a band structure, this time with an energy gap. This distortion to a more stable alternating repeat is the Peierls distortion. It occurs because, when possible, a bond order of 1 is energetically preferred to a bond order of 1/2. The addition of electrons to iodine from a donor may change this situation. However, a powerful donor such as an alkali metal will cause a drastic drop in bond order to form I^- or I_3^- giving the MO situations depicted in Fig. 17 c, and result in insulators as far as electronic conductance is concerned. A weaker donor will transfer a small electron density to I_2, maintain the bond order close to unity, and introduce electrons into the empty antibonding band of poly(I_2). By partial reduction the I_2 chain has become mixed-valent and electrons are delocalized along the chain. Consequent upon a reduction in bond order the I-I spacing will increase and the energy gap between bonding and anti-bonding orbitals decrease. Conductance becomes easier until the stabilizied I_3^- state is reached.

The arguments presented here will apply to many and more complex chains of acceptor molecules, and, in terms of electron deficiency to donors. In organic metals involving CT complexes there is considerable evidence that the introduction of extra aromaticity into a donor or acceptor by electron transfer enhances conductivity[197]; a similar effect is therefore probable in polymeric CT semi-conductors. For transport of charge in organic metals, delocalization and mobility is usually along the crystal stack of donor and/or acceptor moieties. Whether these are the pathways in polymeric CT complexes, or whether the conductance involves migration by alternant electron transfer from donor to acceptor (a mechanism suggested by Eley[198]), or indeed whether both mechanisms occur has not been clearly established.

There are numerous examples in the literature of polymeric CT complexes. Mostly these are formed by CT complexing a polymeric donor with monomeric acceptors. Table 5 shows a limited selection of polymeric CT complexes and their conductivities. CT complexing enhances the conductance relative to the commonly insulating polymeric donor, but if a comparison is drawn between a polymeric CT complex and its monomeric analogue no such simple rule holds. For example the phenothiazine dichlorodicyanoquinone (V, DDQ) system[200] has a $\sigma = 10^{-4}$ ohm$^{-1}\cdot$ cm^{-1}, greater by a factor of 2.5×10^4 than its polymeric analogue [VI in Table 5]. A drop in the conductivity of a polymeric material vis-a-vis a monomeric analogue is not surprising since spatial order is not outstanding in a polymeric phase. Also, local geometric factors have a marked influence on the ability to form CT complexes. Litt and Summers[202] have shown that the polymer (VIII) in Table 5 provides a repeat spacing of the methylmercaptoanisole side chains of 0.635 nm which is typical for most donor-acceptor separations (0.64–0.68 nm). Adding 2,4,5,7-tetranitrofluorenone to the polymer provides a CT polymer whose conductance is greater than its monomer analogue (5×10^{-14} ohm$^{-1}\cdot$ cm^{-1}). The most extensively studied CT semi-conducting polymers are based on poly(vinylpyrididine) (PVP) or poly(N-vinylcarbazole) (VII, PNVC) with a variety of acceptor molecules. The PNVC-I_2 CT system[201] shows variation in σ with I_2 concentration with a maximum $\sigma = 10^{-4} - 10^{-5}$ ohm$^{-1}\cdot$ cm^{-1} at ≈ 70 wt.-% iodine (0.64 mole fraction based on polymer residue molecular weight). The 1 : 1 complex has a slightly lower conductivity, 10^{18} unpaired electrons per gram of which $\approx 10^{14}$ are available as carriers with a mobility of 0.5 cm$^2 \cdot$ V$^{-1} \cdot$ s^{-1}.

It has been reported[203] that complexing PNVC with the polymeric acceptor (X) (Table 5) produces a material with the high ambient conductivity of 10^3 ohm$^{-1} \cdot$ cm^{-1}.

Table 5. Selected polymeric CT complexes and their conductivities

Donor	Acceptor	$\sigma/(\text{ohm}^{-1} \cdot \text{cm}^{-1})$	Ref.
Poly(2-vinyl-pyridine) (PVP)	I$_2$ (75–90% w/w)	10^{-3} a	199)
VI	V(DDQ)	4×10^{-9}	200)
VII (PNVC)	I$_2$ (77% w/w)	10^{-5} a	201)
VIII	IX	10^{-11}	202)
VII (PNVC)	X	10^3	203)

a Varies with I$_2$ concentration; maximum σ quoted.

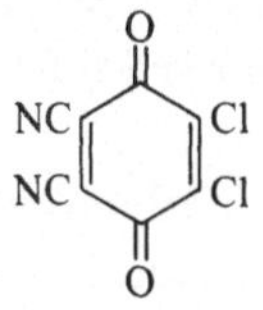

(V)

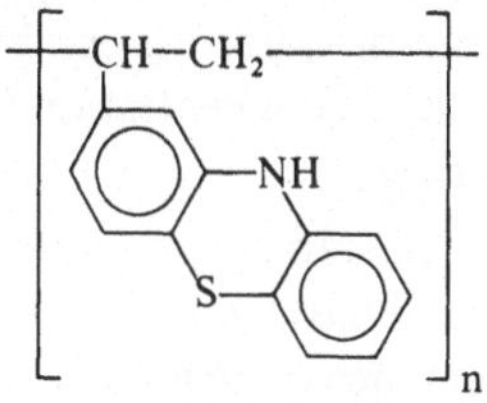

(VI)

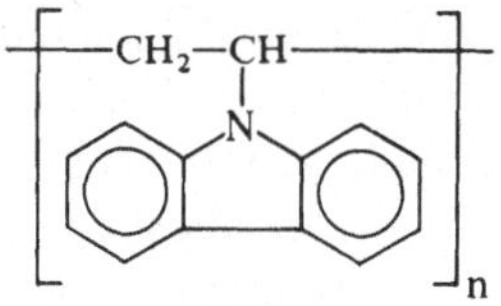

(VII)

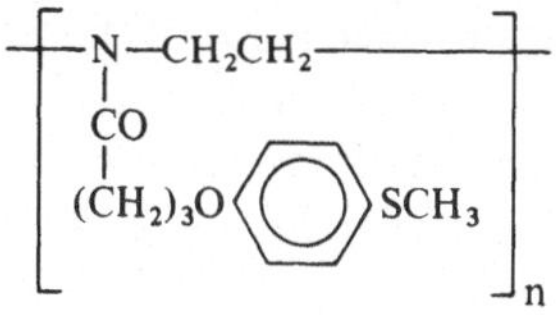

(VIII)

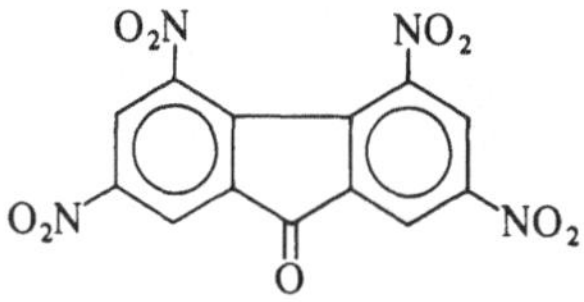

(IX)

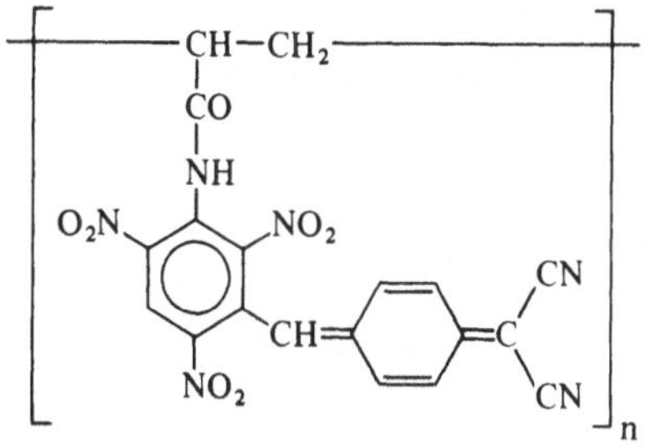

(X)

Other CT polymer-polymer complexes based on polyesters or polycarbonates containing the nitrophthalate moiety as acceptor, and phenyl or anisyliminodiethanol as the donor group in the donor polymer, do not show such a marked enhancement in σ, although σ is increased by $\approx 10^3$ times the homopolymer values[204]. The CT complex of PNVC with 2,4,7-trinitrofluorenone (TNF) has a higher resistivity than the I_2 complex. It has, however, been studied as a photoconductor as has PNVC itself; these aspects are discussed in Sect. 7.1.

As materials most CT polymers are rather brittle, detracting from their potential use as conducting films, etc. They have however, found application in the construction of specialized voltaic cells. The PNVC-I_2 semi-conductor has been used as electrolyte between an active electrode (magnesium, calcium or silver) and an inert electrode (carbon or platinum) to produce a solid-state electrochemical cell. Quite large open circuit voltages (1.5 V for Mg and 2.5 V for Ca) and current densities of 25 mA $\cdot$ cm^{-2} have been reported[205]. The PVP-I_2 complex has found application in long life, miniature batteries made by Wilson Greatbach Limited for implantable pacemakers[199]. The complex acts as cathode against a lithium anode which, during electrolysis provides LiI electrolyte in situ. Open circuit voltages of 2.8 V are reported. The cell has a very high energy density (120 W $\cdot$ h $\cdot$ kg^{-1} as compared to 30 W $\cdot$ h $\cdot$ kg^{-1} for a high quality lead acid battery), a service life of ten years and is relatively light (80 g). The initial composition of the CT complex is 90 wt.-% I_2 which falls during operation. Since the conductivity remains virtually constant at 10^{-3} ohm^{-1} $\cdot$ cm^{-1} in the I_2 concentration range 90–75 wt.-%, the cell functions satisfactorily over this range.

Formally similar to CT semi-conductors are those based on radical ions in which complete electron transfer has taken place between donor and acceptor, but in which enhanced conductance is frequently found when excess donor or acceptor is present. An extensively investigated class of such systems in the polymeric field involve the tetracyanoquinodimethane (TCNQ, XI) molecule and its anion radical (TCNQ$^-$) present as a counterion in a polycation matrix. As a component, TCNQ and its radical ion has provided some of the outstanding organic metals[197] and our knowledge as to the factors promoting good conductance in such systems is of some relevance to any discussion of polymeric systems. TCNQ can undergo a one and two electron transfer as depicted in Eq. (16). The second reduction is energetically much more difficult than the first[206] in which TCNQ$^-$ is stabilized by the introduction of extra aromaticity. In its salt formation, TCNQ$^-$ forms a number of compounds with non-radical organic or inorganic cations D$^+$ (effectively donors) and which may also

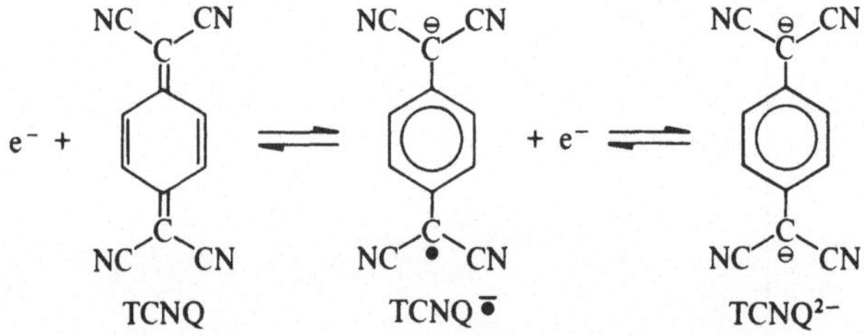

$$ (16) $$

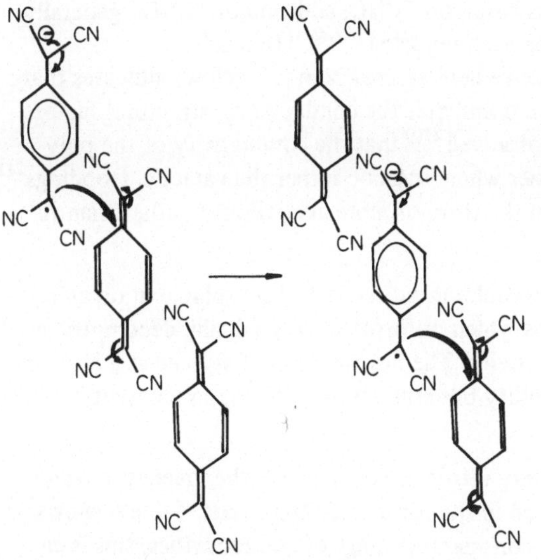

Fig. 18. The stacking of tetracyano-quinodimethane TCNQ and its anion radical TCNQ $\cdot^-$, and the migration of aromaticity in this system

include TCNQ^0: they are of general formulae $D_m^+[(TCNQ)_n]^{\bar{\cdot}}$. Some of the best organic metals are formed when $m = 1$, $n = 2$ and these are often orders of magnitude better conductors than the $1:1$ salts. When this is so, X'ray and other evidence indicates a stacking of TCNQ molecules each of which have a fractional oxidation state, or what is equivalent, form a mixed valent array of TCNQ in which migration of aromaticity is possible (Fig. 18). Spacing, its regularity, and the extent of electron overlap are known to influence the conductivity of the salt, and it is here that the cation size is important. Cations which are planar and approximately the same size as TCNQ generally form the most highly conductive salts of this type. However, the most highly conducting organic metals known do not belong to the series $D_m^+[(TCNQ)_n]^{\bar{\cdot}}$ but rather to $D^{\overset{+}{\cdot}} TCNQ^{\bar{\cdot}}$, that is both anion and cation radicals are involved, and if in the oxidation, aromaticity is also introduced into $D^{\overset{+}{\cdot}}$, conductivity is further enhanced. Thus the TCNQ$^{\bar{\cdot}}$ salt of the 1,4,5,8-tetraselanoful-valene radical cation [XII, Eq. (17)] has a conductivity[207] of 800 ohm$^{-1} \cdot$ cm^{-1}.

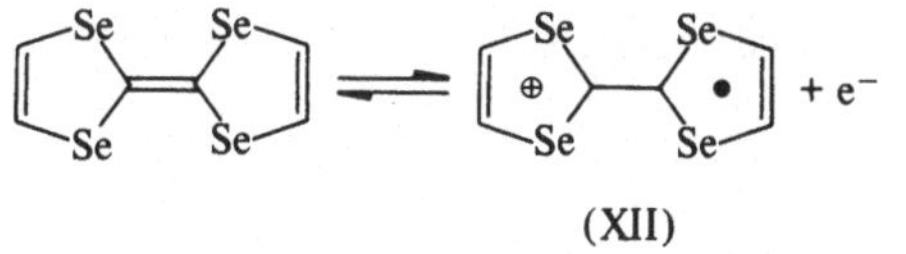

$$\text{(XII)} \qquad \rightleftharpoons \qquad + \, e^- \qquad\qquad (17)$$

A considerable number of polymeric systems involving TCNQ$^{\bar{\cdot}}$ have been synthesised and studied electrically. Mostly, TCNQ$^{\bar{\cdot}}$ is the counterion to quarternized nitrogen. Amongst the polymers so studied are those derived from the vinyl monomers: 2- and 4-vinylpyridine[208–211], 4-dimethylaminostyrene[208], N-vinylimid-azole[208], and a copolymer of styrene with 2-vinylpyridine[208], which are all subsequently fully or partially quarternized. TCNQ$^{\bar{\cdot}}$ is then introduced as counter-ion by double decomposition with Li$^+$TCNQ$^{\bar{\cdot}}$ to produce the polymers. In common with the situation for organic metals based on TCNQ, addition of $TCNQ^0$ generally improves the conductivity which has been reported[162, 211] to reach 1.2×10^{-2} $ohm^{-1} \cdot cm^{-1}$. The evidence here as elsewhere (cf. below) indicates that the dispersed $TCNQ^-$/TCNQ system dominates the conductance; structural factors much less so. Although it has been observed[210] that the conductivity of the polymer based on 2-vinylpyridine is higher when isotactic rather than atactic, Goodings[211] attributes this to long-range order in the form of more crystallinity, rather than to local tactic placements.

These dark coloured materials containing excess TCNQ are soluble in organic solvents and form rather brittle films which unfortunately, gradually decompose in air, subsequently losing their conductivity. The introduction of styrene as a co-monomer[208] yields films of greater flexibility but still with significant conductivity (10^{-3} $ohm^{-1} \cdot cm^{-1}$).

An alternative form of quarternary nitrogen polymers are the ionenes of which many examples have been synthesized in the form of $TCNQ^-$ salts. Table 6 shows some examples. Although there is a considerable range of conductivities, this is enhanced by the addition of $TCNQ^0$. In this respect ionene derivatives parallel the poly(vinylpyridines) and other $TCNO^-$ associated polyanions. Bruce and Herson[217]

Table 6. Conductivities of some ionene-TCNQ systems

Ionene	(x/y)	$[TCNQ^-]/[TCNQ^0]$	$\sigma/(ohm^{-1} \cdot cm^{-1})$	Ref.
XIII	(3/4)	0	3.6×10^{-5}	212)
XIII	(3/4)	0.5	6.7×10^{-4}	212)
XIII	(6/3)	0	$1.05 \times 10^{-5}, 4.8 \times 10^{-7}$	212, 213)
XIII	(6/3)	0.5	$2.9 \times 10^{-4}, 8.3 \times 10^{-3}$	212, 213)
XIII	(6/5)	0	6.7×10^{-9}	212)
XIII	(6/5)	0.5	1.3×10^{-3}	212)
XIII	(6/6)	0	3.1×10^{-9}	212)
XIII	(6/6)	0.5	1.9×10^{-8}	212)
XIII	(6/8)	0	$1.4 \times 10^{-8}, 3.4 \times 10^{-7}$	212, 213)
XIII	(6/8)	0.5	$7 \times 10^{-8}, 5 \times 10^{-3}$	212, 213)
XIII	(6/10)	0	$10^{-7}, 3 \times 10^{-7}$	212, 213)
XIII	(6/10)	0.5	$1.4 \times 10^{-5}, 2.6 \times 10^{-4}$	212, 213)
XIII	(6/16)	0	10^{-7}	212)
XIII	(6/16)	0.5	2.1×10^{-3}	213)
XIV		0	6.7×10^{-7}	214)
		0.5	1.1×10^{-2}	214)
XV		0	1.9×10^{-6}	214, 215)
		0.5	1.3×10^{-2}	214, 215)
XVI		0.5	6×10^{-2}	215)
o-XVII		0	1.8×10^{-6}	216)
		+16% $TCNQ^0$	2×10^{-2}	216)
m-XVII		0	2×10^{-4}	216)
		+16% $TCNQ^0$	1.1×10^{-3}	216)
p-XVII		+15% $TCNQ^0$	9.6×10^{-5}	216)

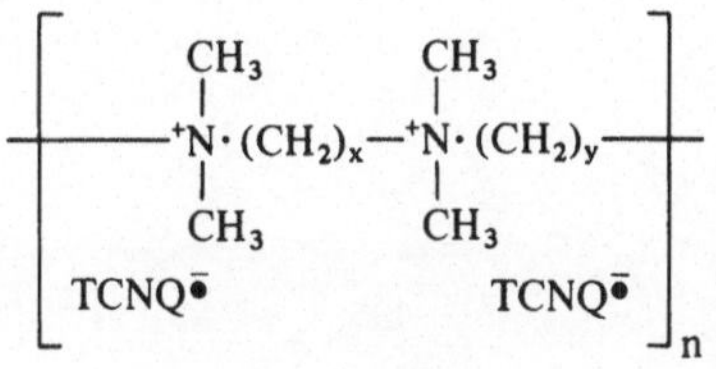

$$\left[\begin{array}{c} \overset{CH_3}{\underset{CH_3}{\overset{|}{-+N\cdot(CH_2)_x}}}-\overset{CH_3}{\underset{CH_3}{\overset{|}{+N\cdot(CH_2)_y}}}- \\ \underset{TCNQ^{\overline{\bullet}}}{} \qquad\qquad \underset{TCNQ^{\overline{\bullet}}}{} \end{array}\right]_n$$

[XIII(x/y)]

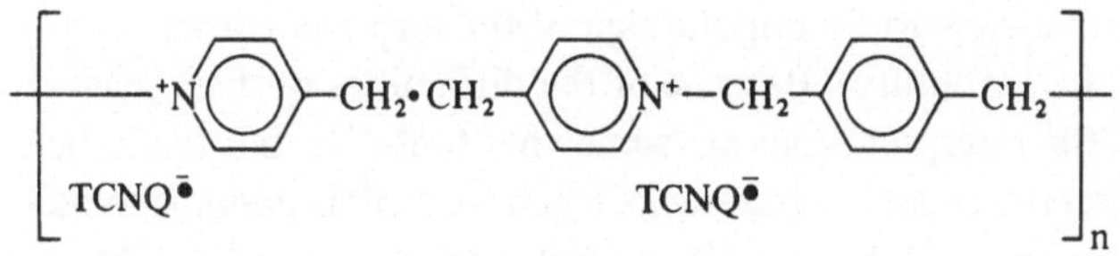

$$\left[\begin{array}{c} +N\!\!-\!\!\bigcirc\!\!-\!\!CH_2\!\cdot\!CH_2\!\!-\!\!\bigcirc\!\!N^+\!\!-\!\!CH_2\!\!-\!\!\bigcirc\!\!-\!\!CH_2\!\!- \\ TCNQ^{\overline{\bullet}} \qquad\qquad TCNQ^{\overline{\bullet}} \end{array}\right]_n$$

(XIV)

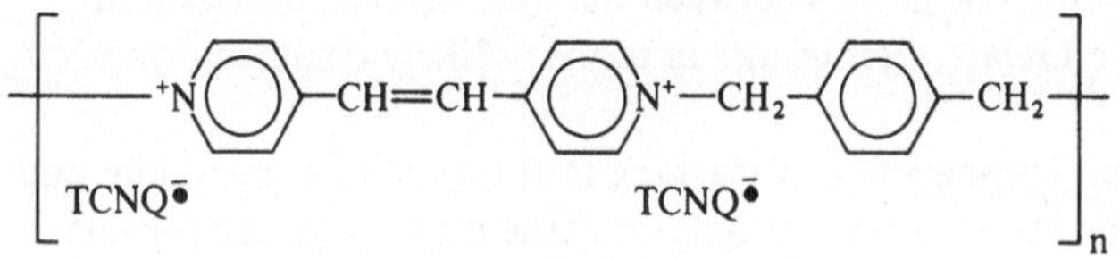

$$\left[\begin{array}{c} +N\!\!-\!\!\bigcirc\!\!-\!\!CH\!\!=\!\!CH\!\!-\!\!\bigcirc\!\!N^+\!\!-\!\!CH_2\!\!-\!\!\bigcirc\!\!-\!\!CH_2\!\!- \\ TCNQ^{\overline{\bullet}} \qquad\qquad TCNQ^{\overline{\bullet}} \end{array}\right]_n$$

(XV)

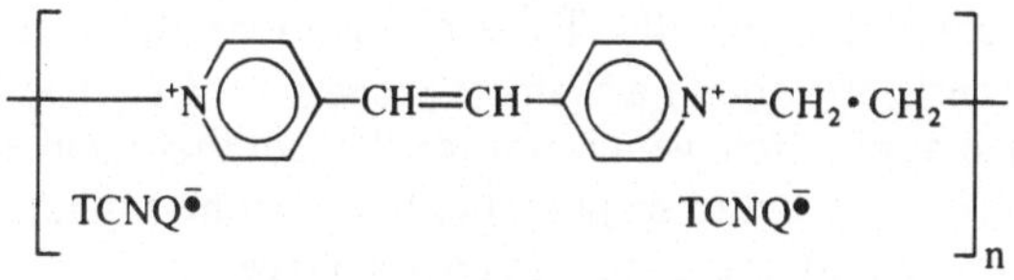

$$\left[\begin{array}{c} +N\!\!-\!\!\bigcirc\!\!-\!\!CH\!\!=\!\!CH\!\!-\!\!\bigcirc\!\!N^+\!\!-\!\!CH_2\!\cdot\!CH_2\!\!- \\ TCNQ^{\overline{\bullet}} \qquad\qquad TCNQ^{\overline{\bullet}} \end{array}\right]_n$$

(XVI)

$$\left[\begin{array}{c} +N\!\!-\!\!\bigcirc\!\!-\!\!\bigcirc\!\!N^+\!\!-\!\!CH_2\!\!-\!\!\left\{\begin{array}{c} o\!-\!C_6H_4\!\cdot\!CH_2\!-\! \\ m\!-\!C_6H_4\!\cdot\!CH_2\!-\! \\ p\!-\!C_6H_4\!\cdot\!CH_2\!-\! \end{array}\right. \\ (TCNQ^{\overline{\bullet}})_{1.5}Br^{\overline{}}_{0.5} \end{array}\right]_n$$

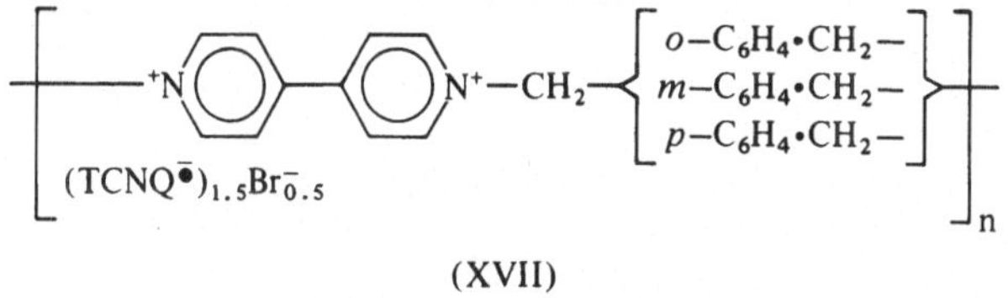

(XVII)

---

have measured this trend over a range of added $TCNQ^0$ to polymer (XVIII) and have found a 1.7×10^5 increase in σ as the mole ratio $TCNQ^0$ to $TCNQ^{\overline{}}$ was changed from 0 to 0.9. Certainly the influence of $TCNQ^0$ is strong support for the view that $TCNQ^{\overline{}}/TCNQ^0$ redox stacks are the paths by which most charge migration occurs in these loaded polymers, and that the polymer itself acts principally as a template for their organization.

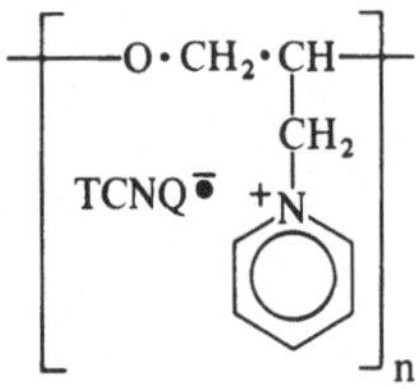

(XVIII)

There have been a number of studies[212, 214)] of Hall effects and measurements of Seebeck coefficients with a view to finding the sign of the majority carrier (and from Hall measurements, their mobility). Because of the difficulties of these measurements with polymers, the interpretations are somewhat tentative, but the indications are that electron transport occurs in the stacks when $TCNQ^0$ is present; in its absence hole conduction appears to take place in at least some members[212)]. The influence of polymer structure is more complex than the effect of $TCNQ^0$ loading, as reference to Table 6 shows. No regular trend of σ values is shown for the ionenes (XIII) as the number of methylene groups between the quarternized nitrogens are changed. Hadek et al. [212)] correlate their results in terms of likely chain conformations.

An interesting apparent consequence of stacking in this type of semi-conducting polymer has come from the observation that polyurethane rubbers containing quarternary nitrogen groups and $TCNQ^-$ plus $TCNQ^0$ [218, 219)] show little change in conductance on extensions[219)] up to 80%. It is suggested that this behaviour, which contrasts markedly with rubbers made conducting (poorly) with carbon black, is due to the stress orientation of stacks enhancing mobility. These observations, the relatively good conductivity, ease of synthesis, and good mechanical properties of $TCNQ^-$ base polymers might well lead to their application were it not for the fact that their stability in air is poor[220)]. This has been found to be particularly so for the aliphatic members. Decomposition leads to a drastic reduction in conductance.

There are far fewer studies of polymers incorporating cation radicals. Manecke and Kautz[221)] have synthesized polymers based on the violene redox systems (XIX) and (XX) which can be oxidized to the cation radical [Eq. (18), $XIX^{+\cdot}$ and $XX^{+\cdot}$] and at higher potential, to the cations.

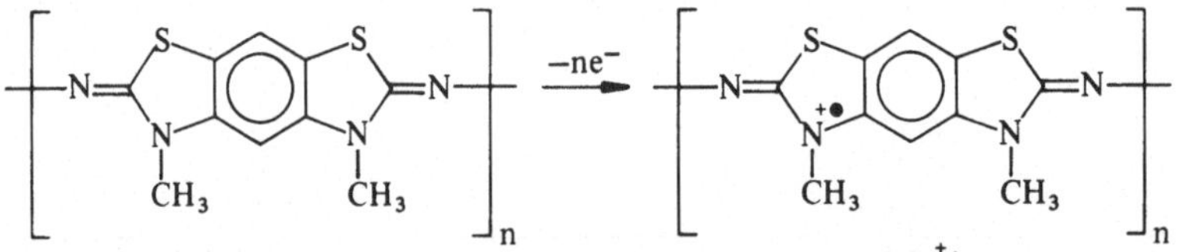

(18)

(XIX)   ($XIX^{+\cdot}$)

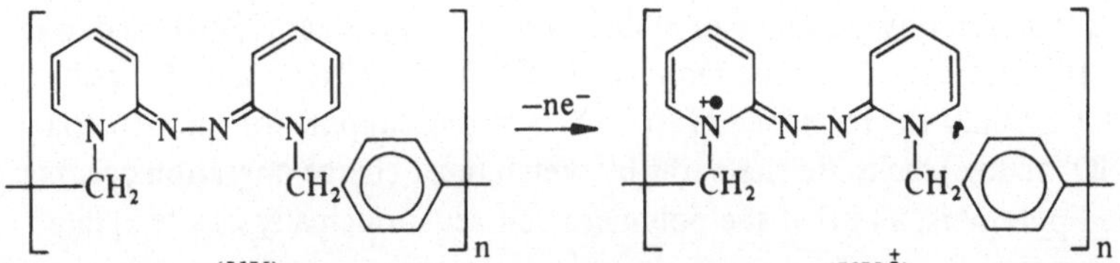

(XX)   ($XX^{+\cdot}$)

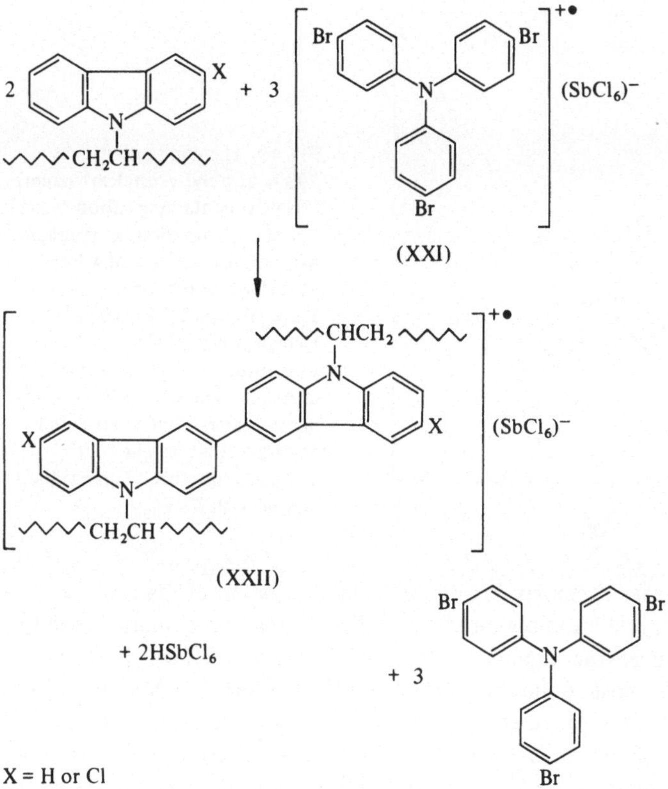

$$(19)$$

X = H or Cl

Some enhanced conductances are shown by the cation-radical forms
(σ_{298}/(ohm$^{-1}\cdot$cm^{-1}); XIX = 1.3 x 10^{-7}, XIX$^{+\cdot}$ = 4.3 x 10^{-5}, XX = 5.5 x 10^{-10},
XX$^{+\cdot}$ = 2 x 10^{-9}, as compressed powders). Cation-radicals have also been introduced
into PNVC by oxidation with tri(p-bromophenyl)ammoniumyl hexachloroanti-
monate (XXI)[222]. The oxidation involves a three electron transfer with concomitant dimerization as illustrated in Eq. (19). The oxidized forms XXII can be compressed to give green glassy discs whose conductivities[5] in the range
10^{-13}–10^{-5} ohm$^{-1}\cdot$cm^{-1} are controlled by the amount of oxidant (Fig. 19). As
Eq. (19) shows, the reaction leads to cross linking of polymer chains which limits
the extent of oxidation before the onset of three dimensional, insoluble networks.
Low molecular weight polymer substrate provides material of highest cation-radical
content and conductance, but, except in limiting the accessible extent of homogeneous reaction, the molecular weight of the substrate polymer does not influence σ. Substitution by chlorine in the carbazole ring also appears to have little
effect on σ. The PNVC cation radicals XXII are, with the exception of those with the
highest radical concentration, stable for at least six months, and even the most highly oxidised material only change σ by one order of magnitude in one month. Removal of the undesirable cross-linking reaction by ring substitution may in future
provide even more highly conducting materials.

---

5   At low oxidation they are also photo-conductive (see p. 156).

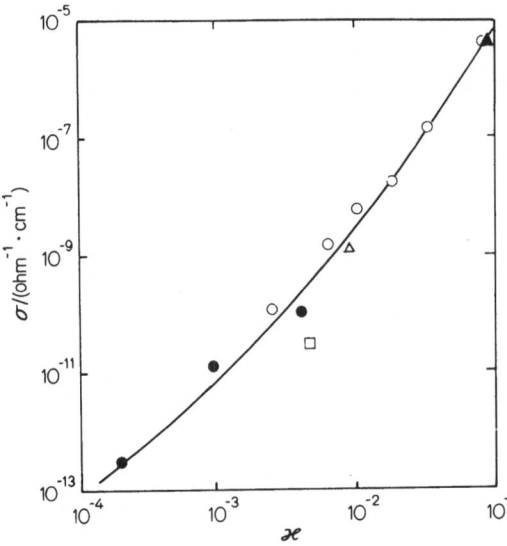

Fig. 19. The conductivities at 293 K of poly(*N*-vinylcarbazoles) (PNVCs) containing cation radical XII at mole fractions *x*. Characteristics of samples: ○ and ● from PNVCs of molecular weights 1.6×10^3 and 1.25×10^6 respectively; ▲, from a PNVC, 75% monosubstituted with chlorine and molecular weight 9.5×10^3; △, a similar polymer ≈90% chlorinated and molecular weight ≈10^5; □, a physical mixture of an oxidised sample with PNVC

There have been recent reports describing the incorporation of the tetrathiofulvene (TTF) heterocycle into polymers[223, 224]. This system, on oxidation readily generates a cation radical [the sulphur analogue of XII of Eq. (17)] which is a component of some of the most conductive organic metals. However, to date the oxidation of TTF$^+$ moieties have provided rather intractable materials or, in one case[224] was unsuccessful with TCNQ or DDQ because of amide deactivation in the TTF containing polyurethane. Hertler[224] has also reported the synthesis of an interesting polymer containing alternant TCNQ and TTF residues via the reaction (20). An intractable material resulted which had a conductivity of 2×10^{-8} ohm$^{-1} \cdot$ cm^{-1}. Addition of I_2 provided a CT complex with ≈1 I_2 per TTF residue and a conductivity of 2×10^{-6} ohm$^{-1} \cdot$ cm^{-1}.

$$ (20) $$

6.3 Organometallic Polymers and Polymers with Metal-Metal Bonds

The introduction of metal atoms into a polymer, or the formation of a polymer through metal centres fequently provides semi-conducting materials. Unfortunately, and equally frequently, the polymers are insoluble, infusable and thus intractable materials. It is beyond the limited space available to discuss the conductance of the large number of known metallo-organic polymers. For this reason a selective approach is adopted here with the aim to illustrate, and whenever possible, to present well characterized systems or ones which have outstanding electrical properties.

Chelation of transition metal ions with a chelate having bis-functionality will produce polymeric co-ordination complexes. A measure of semi-conductance often results when partially filled or vacant d-orbitals overlap π-electron systems and result in increased spatial delocalization. Such polymers are related to the conjugated polymers discussed earlier. An example of such increased conductivity is found in the nickel complex XXIII which has a conductivity (1.4×10^{-10} ohm$^{-1} \cdot$ cm^{-1}) some

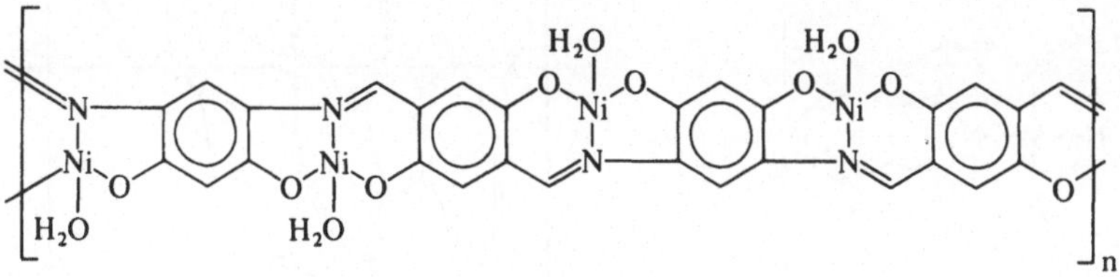

(XXIII)

300 times larger than the uncomplexes polymer[172]. In this example the metal d-orbitals are in cross-conjugation. Polymers in which the conjugation is entirely via metal d-orbitals are known. The Cu^{II} complex of 1,5-diformyl-2,6-dihydroxynaphthalene dioxime is polymeric via co-ordination [structure (XXIV)] and has a con-

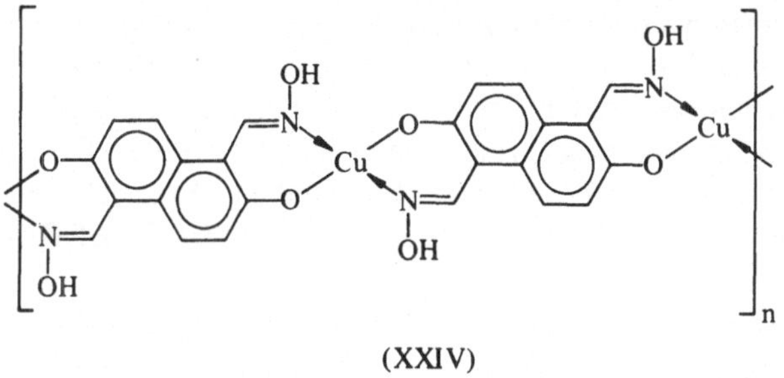

(XXIV)

ductivity[225] of 4×10^{-5} ohm$^{-1} \cdot$ cm^{-1}. With a mixture of transition metals present, conductivities can be further increased[226] suggesting the involvement of mixed valency mechanisms.

Removal or donation of electrons in a mixed valence co-ordination polymer can cause structural change by altering the co-ordination geometry. For optimum conductivity this is likely to be a disadvantage. Square planar complexes in which changes of valency do not result in the redisposition of bonds, avoid this problem. Further, if it can be assured that there is mutual alignment of co-ordinating rings between neighbouring polymer residues so that the d-orbitals are all perpendicular along a polymeric sheet, a high conductance whould result. This principle is the basis of much research into the poly(phthalocyanines)(XXV). Early studies[227] of the Cu^{II} containing polymer indicated good conductivities ($\sigma \approx 10^{-2}$ ohm$^{-1} \cdot$ cm^{-1}) but more recently Norrell et al.[228] have found σ values of 10^{-7}–10^{-4} ohm$^{-1} \cdot$ cm^{-1} for Cu^{I}, Zn^{II}, Sb^{III} and Fe^{III} members, which are lower than the value for the unmetallized polymer ($\sigma_{300} = 1.7 \times 10^{-2}$ ohm$^{-1} \cdot$ cm^{-1}). These workers also observed a strong pressure dependence of conductance.

(XXV)   (XXVI)

Unfortunately, these and other co-ordination polymers having the metal in the backbone of the chain are all too frequently, quite intractable materials. There have been attempts to improve the material properties of poly(phthalocyanines) by incorporating ether or sulphide links, but this has only resulted in a reduction of their conductivities with little if any improvement in tractability[229]. Polymerization of silicone phthalocyanines to produce the stacked ring systems (XXVI) with X = O or X = OC_2H_4O has been accomplished to produce conductive polymers, albeit of lower conductivity ($\sigma = 6 \times 10^{-7}$ ohm$^{-1} \cdot$ cm^{-1} for X = O and 4×10^{-11} for X = OC_2H_4O) than the planar polymerized variety[230].

The moderately good conductivity and high thermal stability of poly(phthalocyanines) and similar polymers has led to efforts to devise fabrication procedures, notwithstanding their non-moldability. Methods of in situ formation of porphyrin-like metal polymers by heating metals with multi functional nitriles, or by direct deposition onto metal surfaces by gas phase reactions have been described[231].

Organo-metallic polymers based on sandwich compounds provide another group of substrates which can provide semi-conductors. Simple non-polymeric ferrocenes (XXVII), biferrocenes (XXVIII) and biferrocenylenes (XXIX) form a class of organic metal with TCNQ[232] which involve the mixed Fe(II)/Fe(III) valencies. Several polymeric analogues have also been synthesized and their electrical properties measured.

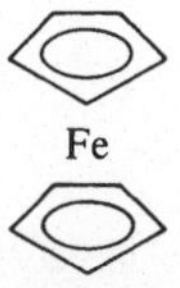

(XXVII) 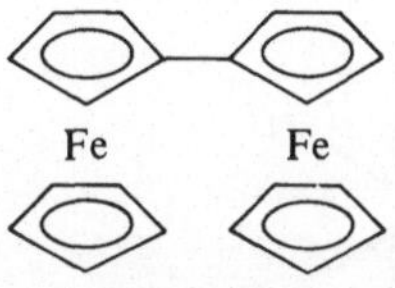　　　　　(XXVIII) 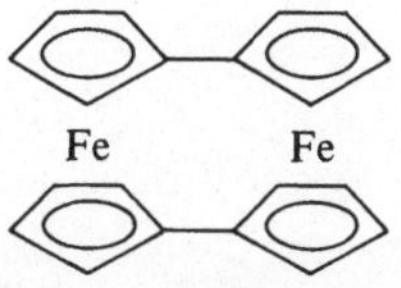　　　　　　　　(XXIX)

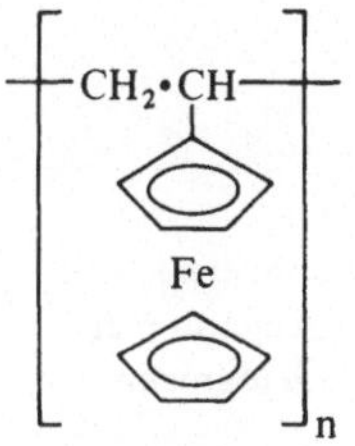

Fe　　　　　Fe　　　Fe　　　　　　Fe　　　Fe

$$\left[CH_2 \bullet CH \right]_n$$ (XXX)

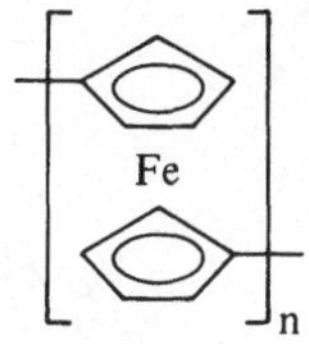

Fe

(XXXI)

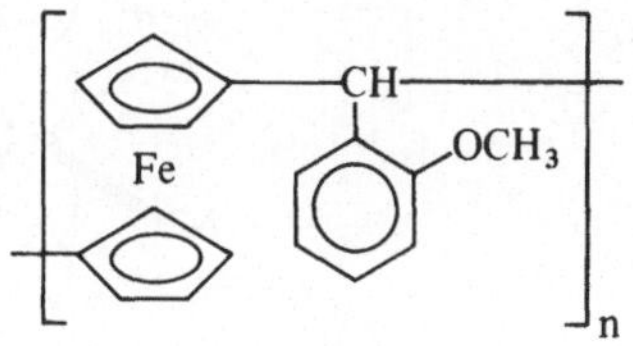

(XXXII)

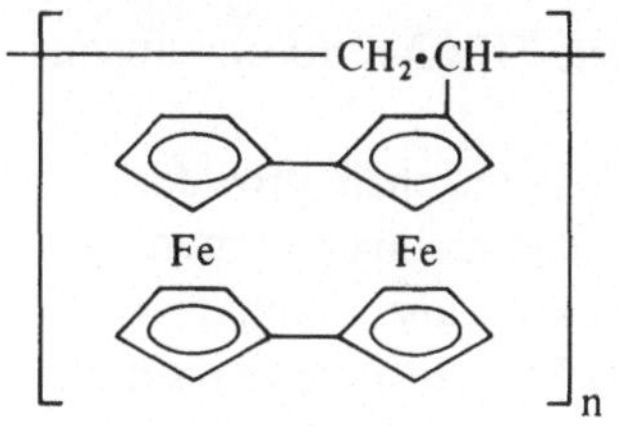

$$OCH_3$$

(XXXIII) 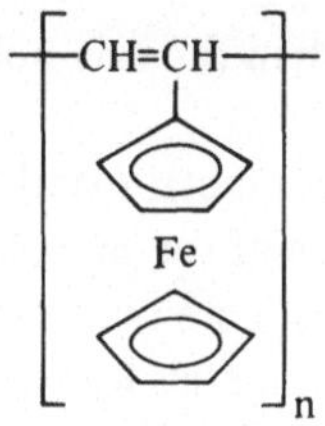　　　　　(XXXIV) 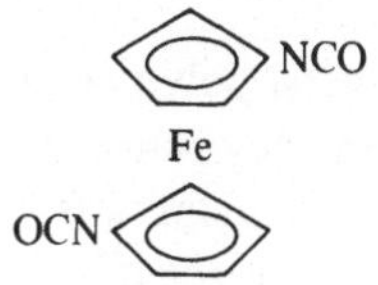　　　　　(XXXV)

$$CH_2 \bullet CH$$　　　$$CH=CH$$　　　NCO

Fe　Fe　　　　Fe　　　Fe

OCN

The polymer **XXX** in the Fe(II) state has a low conductivity
($10^{-13} - 10^{-15}$ ohm$^{-1} \cdot$ cm^{-1}) but subsequent to partial oxidation with picric acid,
benzoquinone, chloranil or TCNQ produces intensely coloured salts with maximum
conductivities at 70% oxidation of $\approx 10^7$ times that of the parent[233]. Such increases
also occur with poly(ferrocenylene) (**XXXI**)[234], polymer (**XXXII**)[235], poly(ethynyl-
ferrocene) (**XXXIV**)[236] and poly(3-vinylbisfulvalenedi-iron)[237] (**XXXIII**) upon partial oxidation. Polymer **XXXIV** incorporates both a conjugated π and redox systems.
It was only of low molecular weight due to the difficulties of free-radical polymerizing the alkyne monomer. Some instability in solution was noted which was not
found in the solid state. Polymer **XXXIII** based on the fulvalene di-iron system is
after 71% oxidation with TCNQ, the most conductive sandwich polymer known at
present ($\sigma \approx 10^{-2}$ ohm^{-1} cm^{-1}). This is in keeping with the fact that organic metals
based on a fulvalene di-iron tend to have the highest conductivities of the group
(**XXVII–XXIX**). High conductance in (**XXXIII**) is accompanied by intractability;
it is reported to be a refractory powder, and efforts to mitigate this by copolymerization led to a reduction in σ[236]. Hertler[224] has recently reported the synthesis of
1,1'-diisocyanatoferrrocene (**XXXV**) and its polyurethane formation with the TCNQ
substituted diol shown in reaction (20). This alternating donor-acceptor electron
transfer system had a conductivity (as a compacted powder) of 3.3×10^{-4} ohm$^{-1} \cdot$ cm^{-1},

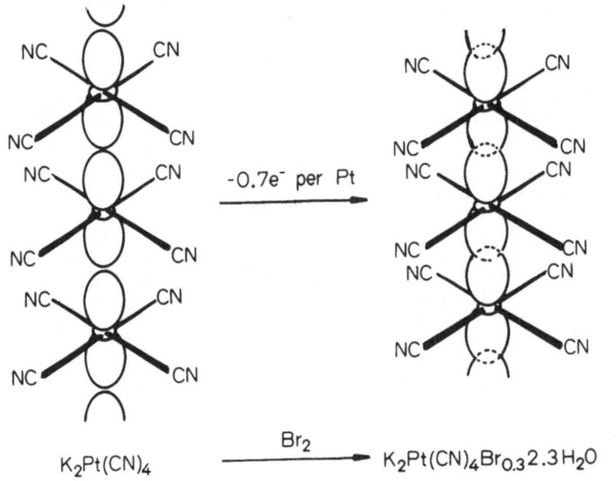

$-0.7 e^-$ per Pt

$K_2Pt(CN)_4$ $\xrightarrow{\quad Br_2 \quad}$ $K_2Pt(CN)_4Br_{0.3}2.3H_2O$

Fig. 20. The conversion of potassium tetracyano platinate(2−) to a Krogmann salt

some two orders of magnitude lower than the non-polymeric TCNQ-ferrocene organic metal.

Transition-metal polymers involving metal-metal bonds in the chain provide, when in mixed valency states, examples of the most highly conducting polymeric systems. These Krogmann salts[237] which are nominally square planar complexes (usually with CN⁻) of the d^8-transition metals (particularly Pt or Ir) involve d-orbital overlap in chain bonding as shown in Fig. 20. Removal of some electrons from d_{z^2} antibonding orbitals of the precursor $[Pt(CN)_4]^{2-}$ induces $d_{z^2}-d_{z^2}$ overlap by removing electrons from filled d_{z^2}-orbitals, reduces the Pt-Pt separation, and renders the oxidation state of platinum non-integral. A very good one-dimensional conductor results[238-239]. Optimum conditions are found at an oxidation state for Pt of 2.3 which leads to the formation of a conduction band (based on d_{z^2} overlap) 5/6 full and a Pt-Pt separation of 0.289 nm rather than the o.335 nm for the non-oxidized Pt(II) form. The material, now of formula $K_2Pt(CN)_4Br_{0.3}$ 2.3 H_2O rather than $K_2Pt(CN)_4$, forms crystals which are optically and electrically highly anisotropic[238] $(\sigma_{||} \approx 10^2$ ohm$^{-1} \cdot$ cm^{-1}; $\sigma_{||}/\sigma_{|} \approx 10^5$). In temperature dependence $K_2Pt(CN)_4Br_{0.3}2.3 H_2O$ behaves as a one-dimensional metal down to 120 K whereupon there is a phase change which is accompanied by a metal-to-insulator transition[240]. Thus at lower temperature the conductivity decreases to $\approx 10^{-12}$ ohm$^{-1} \cdot$ cm^{-1} at 20 K: the hopes for superconduction where thus disappointed! As solids the Krogmann salts have poor mechanical properties being very brittle.

6.4 Intercalated Conducting Polymers

It has been known for many years that the addition compounds between graphite and alkali metals or bromine result in enhanced conductivity[241]. Recently AsF₅ has been found to be particularly effective in this regard[242]. These materials appear to act as donor or acceptor molecules to the conjugated π-system which is "ampho-electronic" in donating or accepting electrons. Very recently there have been reports[243–245] that *cis*- and *trans*-polyacetylene also have their already significant conductivities (p. 131) markedly increased by doping with alkali metals, halogens or AsF$_5$. Conductivities approaching 10^3 ohm$^{-1} \cdot$ cm^{-1} have been reported for micro-crystalline specimens. Experimentally, there is a finite, but very low activation energy which, by analogy to the behaviour of microcrystalline (SN)$_n$ gives grounds for believing that individual crystals are metallic. Spectral evidence also supports this view.

It has been stated[245] that an aligned, fibrous sample doped with AsF$_5$ has an anisotropic conductance with a conductivity in the direction of alignment of 2,000 ohm$^{-1} \cdot$ cm^{-1}. The AsF$_5$ and I$_2$ doped materials have been reported to be stable during one week in vacuum measurement, although air causes decomposition and a decrease in σ taking a period of days[244]. As well as these interesting observations, doping by donor or acceptor leads respectively to n or p type conductance, and a sandwich material has been shown to act as a rectifier. The high conductivity, control of conductance (over 12 orders of magnitude) and of carrier sign certainly make these promising materials, whose development is awaited with interest.

7 Photoconductivity in Polymers

The investigation of photoconductance in polymers has become a fluorishing topic of research and development. The reasons for this are two fold: photoconductance data on polymers (as on other classical photoconductors) provide a fruitful source for understanding the solid state physics of such materials, and their practical exploitation has great potential. To be considered under the latter are the practical development of devices for abstracting solar energy, the attainment of commercially viable electro-imaging machines based on polymers, as well as the application of polymers as the photo active elements of a whole gamut of photoelectric devices. Electro-imaging using a PNVC based photoconductive element is used in a commercial photo-copying machine. Much research into PNVC and its doped form with CT acceptors, particularly 2,4,7-trinitrofluorenone (TNF) has been undertaken. The motivation to commercially exploit its properties is evident[246, 247].

The technique of electro-imaging or xerography for which PNVC/TNF or selenium are the photo active elements is illustrated in Fig. 21. A corona discharge charges a photo-conductive film which will only significantly discharge under illumination. The manuscript to be copied is illuminated, and light reflected from the print free regions of the surface depolarize these regions of the charged photo-conductive film. In consequence this now has a latent image of charge corresponding to the manuscript's text. Resin encapsulated carbon particles which have been charged (usually self charged triboelectrically) are then introduced, and adhere electrostatically to the latent image. Transfer to paper is accomplished by charging the paper. The image is set on the paper by heating so as to fuse the resin skin of the pigment.

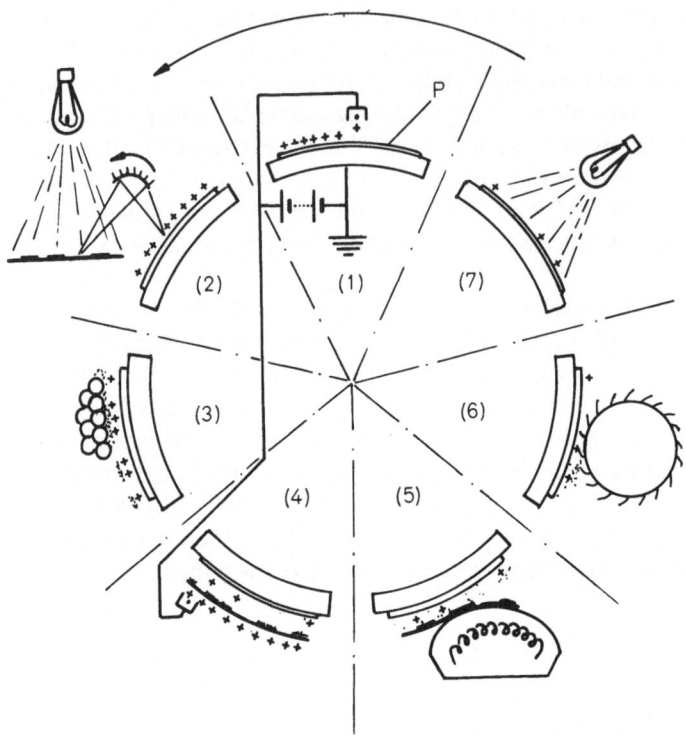

Fig. 21. Schematic illustration of the steps in xerography. (*1*) Surface charging of photoconductive plate P. (*2*) Copying the document by scanning with light, thereby selectively discharging the plate. (*3*) Selective addition of toner onto charged areas by carrier beads. (*4*) Electrostatic transfer of toner onto paper. (*5*) Heat setting of image. (*6*) Removal of excess toner from plate. (*7*) Re-illumination [same lamp as in (*1*)] to remove remaining charge on plate

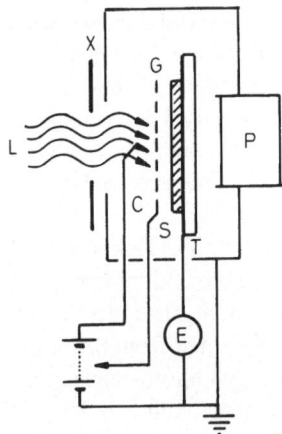

Fig. 22. The photo-induced discharge technique. *C*, Corona discharge electrode with grid *G* to spread an even surface charge on sample *S*. Monochromatic light *L* controlled by a shutter *X* falls on the sample which discharges via the semi-transparent rear electrode *T*. The rate of discharge is recorded by an electrometer *E*. *P* is a photo-detector which monitors the light intensity

Mechanical removal of excess pigment from the photo-conductive ram, and its total depolarization by illumination completes the process.

Components of this sequence form the basis of the 'photo-induced discharge' technique of studying photo-conductance which is particularly well suited to polymers. A corona discharge is used to charge a polymer film and its subsequent loss of charge upon illumination monitored by means of an electrometer (Fig. 22). The use of selenium as a coating on a polymer film so as to provide a well characterized source of photo-generated carriers extends this technique to the study of semi-conductance referred to on p. 128. Extension to 'time of flight' measurements is made possible by using a pulse of light. With the more conductive photo-conductors the direct method of resistance measurement as used for semi-conductors (see p. 127) can be applied, provided a semi-transparent electrode is used.

As indicated earlier, studies of photoconductance can provide much information on the electronic states in solids and on the mobility of carriers. Many books and articles are available on this topic: in the opinion of this writer those listed[248-251] are particlularly germane to this article. Briefly, the salient features of photoconductance are as follows. The photophysical generation of carriers can occur in a number of ways as illustrated in Fig. 23. Firstly, the metal electrode may itself be the photo-active agent and photo-inject either electrons into the conduction band (ia) or receive photo-injected holes from the valence band (ib), processes deliberately chosen in the previously described 'photo-induced discharge' technique employing selenium coating. Route (ii) involves the direct promotion of electrons into a conduction band. Generally such transitions require the high energies of the vacuum ultra-violet, X'rays γ-rays or electron beams, and have not been unequivocally shown to occur in polymers. In photoconductance involving excitation in the sample by visible or ultra-violet radiation, the primary step entails populating some non-conducting electronic state [route (iii)]. For conduction to take place the generated exciton requires further energy to promote an electron into the conductance band (step iv). A number of mechanism are known by which this is achieved. Exciton migration through the

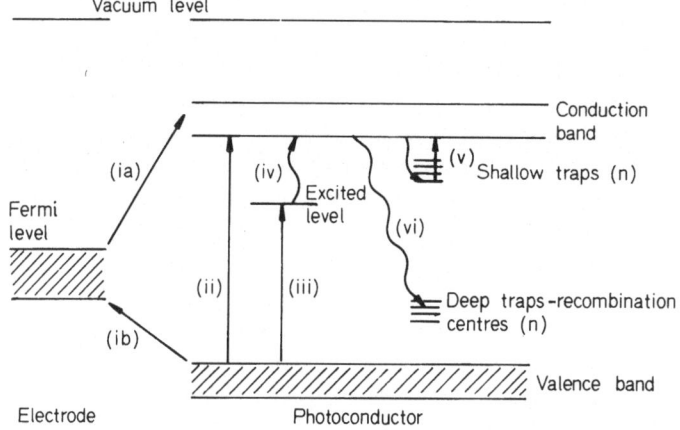

Fig. 23. The generation and trapping of carriers; for description of the various processes see the text

sample to extra energy sources at the surface; migration to allow exciton-exciton interactions; photo stimulation of the exciton by further light absorption, and even direct two-photon interactions. Frequently characteristic of routes (iii) plus (iv) is some mirroring of the light frequency dependence of photoconductance with the absorption spectrum. Photo-detrapping from trapped states (mechanism v), although strictly not a primary method of generating carriers, since the traps must previously have been populated, is certainly a common phenomenon, particularly in polymers. Characteristics of the significant involvement of traps are an irreproducibility of behaviour, a strong dependence of conductance on previous history, particularly in regard to its exposure to light, and sluggish response to illumination.

The photocurrent stimulated is limited in all of these processes by the rate of carrier production (light intensity dependent) or by the carrier mobilities, or both. Systems in which the current is limited entirely by the rate of carrier generation are termed emission limited, whilst if carrier mobility determines the current it is space-charge limited (SCLC). Indicative of the onset of SCLC is supralinearity in the current-voltage dependence: theoretically I then varies as V^2.

The presence of traps has a very marked effect on photoconductance and they are prevelant in most polymers both as structural defects and extraneous impurities. Doping of polymers to alter the photophysical and other characteristics of polymers may also introduce more such traps. Traps are conveniently, but somewhat arbitrarily, classified as deep or shallow in relation to the ease with which re-promotion into the conductance band takes place. It is more difficult to excite a charge trapped in a deep trap and in consequence its probability of release is less.

The formation of a carrier (electron or hole) results in an oppositely charged partner which may be mobile (resulting in electron plus hole conduction) or trapped. If it is trapped, or becomes so subsequently, it can capture mobile carriers of opposite charge and in some circumstances permanently remove them from the supply of carriers. When it does so, it is termed a recombination centre. Deep traps for electrons are by their energetic position more accessible to holes and vice versa, and it is deep traps which are the source of recombination centres [Fig. 23(vi)].

The significance of traps in photoconductors, particularly polymeric ones can be appreciated by considering the nature of photoconductance in their absence. Electrons or holes in conductance or valence bands, have high mobilities and the response of a field would be very rapid and the transit time between the electrodes would be very short. Such is not the case in most materials, particularly not in polymers where mobilities are low. The reason is that the capture by and re-emission from traps prolongs the transit through the sample. Further, the response to a light pulse would show a time lag, sometimes of seconds or minutes when a high density of traps or recombination centres are present. Space forbids development of the theory underlying these effects and the reader is referred to the treatment given by Rose[248]. However, Fig. 24 shows how the presence and proportions of traps and recombination centres can effect the response of photocurrent. This type of experimental data, together with light intensity (J), T and E dependence provide the information for investigating the solid state physics of photoconductors. Variations with wave length (λ) and with contact metals of different work function provide further insight into such phenomena.

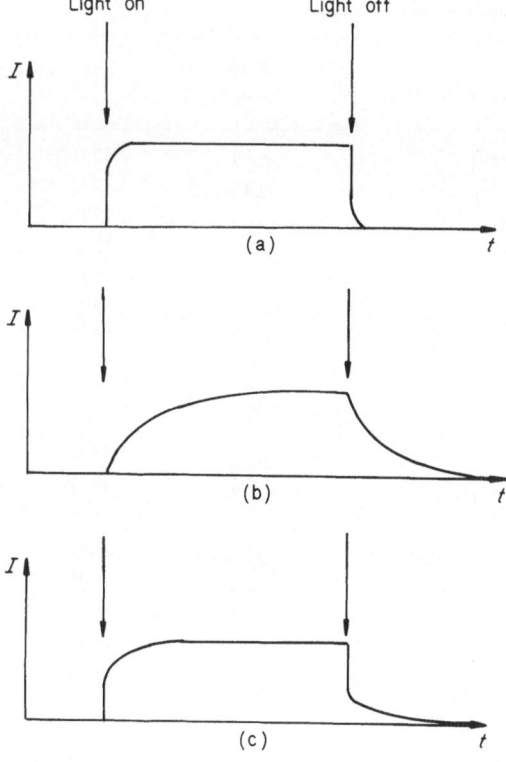

Fig. 24a–c. The photo current (I) response of a photoconductor under an 'on' and then 'off' sequence of illumination. a Behaviour with a high density of recombination centres (or low light intensity) and no traps. b As (a) but with traps present whose population is rapidly equilibrated in relation to the life-time for recombinations. c As (b) but with a slow release from traps relative to the life-time for recombinations

7.1 The Photoconductance of Polymers Based on N-Vinylcarbazole

Photoconductance studies have been reported on a number of polymers (Table 7) including materials not generally associated with photoconduction, thereby exemplifying the use of photoconductance studies in solid state physics. PNVC alone, and with additives has received most attention and will be discussed prior to reviewing a number of other photoconducting polymers. This polymer was first reported by Hoegl[205] to be photoconductive. It responds with a high gain, to ultra-violet radiation below 500 nm, and can be sensitized with CT acceptors or dye molecules.

Hole transport has been established for undoped PNVC by a number of mobility studies using transit time measurements[259–262]. Most investigators[259–262, 266] report a low, field dependent mobility ($\mu \approx 10^{-7} \, cm^2 \cdot V^{-1} \cdot s^{-1}$ at $10^5 \, V \cdot cm^{-1}$) although Szymanski and Labes[261] found a higher, field independent mobility ($\mu \approx 10^{-3} \, cm^2 V^{-1} s^{-1}$). Photo-injection of electrons into deposited electrodes also confirms hole transport[259, 262, 269]. The threshold energies for photo-emission and work functions of the metals (Se[259]; Au, Cu, Al[255, 262, 269]) establishes the edge of the valence bands as 6.1 ± 0.4 eV relative to vacuum. From the fine structure of the quantum yield spectrum the width of the valence band is $< 0.1 \, eV$[269]. The energetics of the processes occurring in PNVC (and PNVC/TNF) have been schematically illustrated by Mort and Emerald[267] and are shown diagrammatically in Fig. 25.

Table 7. Characteristics of some photoconducting polymers

Polymer[a]	Carrier [mobility μ/cm$^2 \cdot$ V$^{-1} \cdot$ s^{-1})][b]	Activation energy/eV	Ref.
Poly(vinyl acetate)	p$[10^{-4}-10^{-5}]$	0.62	252)
	? $[2.2 \times 10^{-12}]$	0.48/1.2[c]	252)
Poly(vinyl chloride)	n$[7 \times 10^{-4}]$	–	252)
	n$[5 \times 10^{-4}]$	–	254)
Polyethylene	p$[10^{-10}]$	–	255)
	? $[10^{-7}]$	–	256)
	n$[10^{-3}]$	0.06	257)
'Polythene'	n$[10^{-10}-10^{-7}]$	0.7–1.2	256)
Poly(ethylene terephthalate)	n$[10^{-6}]$	0.2–0.35	255)
	n$[1.2 \times 10^{-4}]$ + p $[2.7 \times 10^{-5}]$	0.3–0.32	258)
Poly(methyl methacrylate)	? $[2.5 \times 10^{-11}]$	0.52	253)
'Perspex'	? $[3.6 \times 10^{-11}]$	0.48	253)
Poly(butyl methacrylate)	? $[2.5 \times 10^{-10}]$	0.65	253)
Polystyrene	p$[10^{-6}]$	0.2–0.75	255)
	? $[1.4 \times 10^{-11}]$	0.69	253)
Polyisoprene	? $[2 \times 10^{-12}]$	1.10	253)
'Lucite'	? $[3.5 \times 10^{-9}]$	0.52	253)
PNVC	p$[10^{-6}-10^{-7}]$	0.4–0.7	259)
	p$[10^{-6}]$	0.36	260)
	p$[10^{-3}]$	0.12	261)
	p$[10^{-6}]$	–	262)
PNVC/I$_2$	? $[0.5]$	–	263)
	p$[0.1]$ + n$[0.1]$	–	264)
PNVC/TCNE	n$[2 \times 10^{-2}]$ + p$[2 \times 10^{-2}]$	–	264)
PNVC/TNF (1:1)	n$[10^{-5}-10^{-7}]$	–	265)
PNVC/TNF (various)	n$[10^{-6}-10^{-8}]$ + p$[10^{-5}-10^{-8}]$	0.65–0.71	266)
	n$[10^{-6}-10^{-10}]$ + p$[10^{-6}-10^{-8}]$	–	267, 268)

[a] PNVC = poly(N-vinylcarbazole); TCNE = tetracyanoethylene; TNF = 2,4,7-trinitrofluorenone.
[b] p = hole, n = electron and ? = undetermined carrier.
[c] Above and below glass-rubber transition temperature.

Early in the study of the photoconductivity of PNVC it was appreciated that it would be advantageous to extend the wave length response into the visible. In his original observations, Hoegl[205, 247] demonstrated that low levels (<0.10 mole) of acceptor molecules were effective in this respect. Since these observations many studies of CT complexed PNVCs have been undertaken particularly with TNF, I$_2$ and tetracyanoethylene (TCNE). CT complexes with I$_2$ or TCNE have been investigated in some detail[201, 213, 264, 270]. At a 0.04 mole level of acceptor, drift mobilities of ≈ 0.02 cm$^2 \cdot$ V$^{-1} \cdot$ s^{-1} and 0.1 cm$^2 \cdot$ V$^{-1} \cdot$ s^{-1} were observed for both holes and electrons, which are much higher than other CT complexes of PNVC[264]. Significant semi-conductivities are also observed. These observations, particularly those involving an I$_2$, are consistent with transport along acceptor chains involving an I$_2$/I$_3^-$

redox system. In application, particularly for xerography the levels of semi-conductance detract from the value of I_2 or TCNE doped PNCVs. The PNVC/I_2 complex also produces a photovoltaic effect: a 75 μm thick sample of 6.3 cm^2 areas when illuminated with light of intensity 20 mW $\cdot$ cm^{-2} produced photovoltages of some 5–15 mV depending on λ[201].

TNF enhances and extends the wave length range of photo-conductance without any dramatic increase of semi-conductivity, and in consequence its potential in xerography has led to this system being the most extensively studied. The phase compatibility of PNVC to TNF is also superior to most other acceptor molecules: loadings up to 1.2 moles of TNF per residue mole of carbazole are possible as against 0.1 to 0.2 for other acceptors. The ability to easily cast homogeneous films from solutions of the two components makes this system ideal for study. A spectroscopic investigation by Weiser[271, 272] has established much of the detail of CT complexing of PNVC in solution and the solid state. Two CT absorption bands occur at 2.17 and 2.83 eV in both film and solution. The extent of association in films between TNF and carbazole residues varies from 71.8% associated acceptor, at mole ratio TNF/carbazole of 0.1 to 39.2% at a mole ratio of 1.1. The concentration of the CT complex increases only slightly at mole ratio >0.4. Weiser[271] has also reported an interesting influence of electrostatic field on the absorption spectrum of PNVC/TNF with an increase of absorption coefficient of the CT bands with E (Stark effect). Analysis indicates that the effect is too large to be due to changes of polarizability and is probably due to an induced dipole resulting from electron exchange on excitation. Photogeneration of radicals has been demonstrated in the PNVC/TNF system by irradiating the CT band at low temperatures (<−40 °C) to give a structureless e.s.r. spectrum[273] assigned to TNF$^{\bar{}}$. The signals increased photoconductances. All these observations are consistent with at least part of the conduction mechanism involving the sequence: CT complexing, electron promotion by light adsorbtion, and thermal formation of TNF$^{\bar{}}$ as its complex with neighbouring TNF molecules. A schematic representation of the photophysical processes are illustrated in Fig. 25.

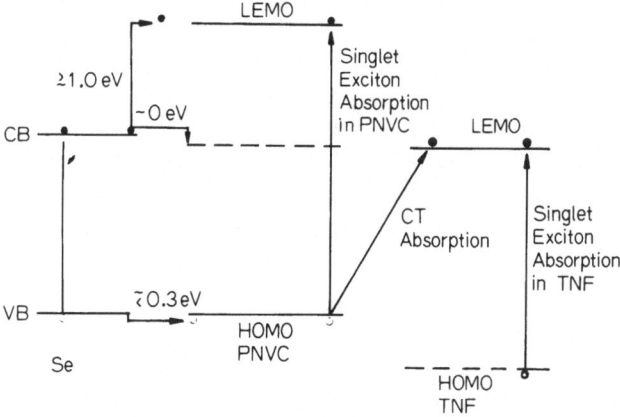

Fig. 25. The relative energy levels and processes occurring in amorphous selenium and the PNVC/TNF system (from Mort and Emerald[267]). HOMO, highest occupied molecular orbital; LEMO lowest empty molecular orbital.●, Electron; ○, hole

The transport phenomenon occurring in PNVC/TNF films has been studied by a number of workers[265–268, 274–276]. Schaffert[274] first reported that in contrast to PNVC, PNVC/TNF at high TNF loadings showed electrons as majority carrier. Gill[266] has demonstrated that both hole and electron transport can occur in PNVC/TNF system, and that the hole conductance is associated with the PNVC phase whilst electrons transport proceeds via the TNF molecular arrays. He further showed that hole and electron mobilities are both E and T dependent with $\mu \propto \exp\{\gamma E^{1/2}\}$. It was observed that γ did not depend on carrier sign or PNVC/TNF ratio. Measurements on pure TNF as a single crystal; in the liquid state; as a glass; or dispersed in polyester, showed the same form of E dependence for the last two systems[276]. These and other observations fit a Poole-Frenkel mechanism[277, 278] in which there is trap-controlled conductance. A field lowers the escape energy from traps by decreasing the depth of the Coulomb potential well in the direction of the field. Such a model predicts an $E^{1/2}$ dependence for mobility. Other models have been suggested[279] but do not fit the data, (particularly that of Fox[275]) so well. Gill[278] has also reported on measurements of trap depth as a function of E and T using thermally stimulated current measurements. The results are entirely consistent with the mobility studies. Together the measurements of electron mobility as a function of TNF concentration in doped PNVC and polyester[276], and optical absorption experiments[271] indicate that all the TNF molecules, and not just those that are complexed, are involved in the transport of electrons. For hole transport carbazole residues provide the pathway, but whether in TNF doped material, complexed carbazole residues sustain hole transport is not so well established. Gill[266], on the basis of a dependence of μ_+ on the average separation of uncomplexed carbazole residues, rather than total carbazole density, provides some evidence in favour of the non-involvement of complexed carbazole in hole transport.

The sensitization to visible radiation of the photoconductance of PNVC has also been achieved by the addition of dyes such as crystal violet, benzopyrylium or carbonium salts[280]. Little is known of the detailed mechanism of such 'optical sensitization'. Mylnikov and Terenin[281] have suggested that the dye molecules provide sites from which photo-injection of electrons occurs, or, alternatively, that these transfer energy after light absorption to release trapped carriers from trapping levels.

Recently, PNVC oxidised to a low level of cation radical (mole fraction $\leqslant 4 \times 10^{-3}$) has been shown to be photoconducting[282]. The response is panchromatic over the wave length range 400–600 nm. In response, these materials show considerable time lags, consistent with the participation of traps. Both activation energy measurements and the differing dependence of photo-current on light flux at 20 °C and −20 °C indicate that some conformational or phase change may occur in this temperature range. The nature of the carriers has not been established.

Pittmann and Grube[283] investigated the electrical and photoelectric properties of copoly(N-vinylcarbazole, vinylferrocene) in the presence of TNF and in which Fe(II) was partial oxidised with DDQ. They found by independent oxidation experiments that TNF does not complex or oxidise the ferrocene moiety nor DDQ the carbazole rings. The semi-conductivity was due to a mixed Fe(II)/Fe(III) valence system. As a photoconductor, the copolymer was less active than PNVC/TNF systems of similar loading. Other copolymers of N-vinylcarbazole and vinyl monomers have been investigated for their photoconductive behaviour[284]. Most copolymers after TNF addition appear to be less effective than PNVC/TNF as photoconductors. The nitration of PNVC is reported[285, 286] to enhance its photoconductance. Here CT complexing between carbazole and nitrated carbazole residues is the probable cause.

An interesting copolymer system has been described by Okamoto et al.[285]. This involved a copolymer of N-vinylcarbazole with 4"-vinyl Malachite Green to produce a copolymer with in-chain dye sensitization. Corona discharge techniques established the copolymer was a less effective photoconductor than PNVC doped with malachite green, but more effective than a physical blend of PNVC with poly(4"-vinyl malachite green). Increased trapping in the copolymer and aggregation of malachite green residues in the blend were stated to be the cause for this order of response.

7.2 The Photoconductance of Polymers Not Based on N-Vinylcarbazole

Polymers other than those based on N-vinylcarbazole have also been studied as photo-conductors, although not so extensively. Williams et al.[287] have found hole transport in poly(2-vinyl-N-alkylcarbazoles) and the 3-vinyl analogues. They also measured the energy levels involved. Of the common, industrial polymers few show any promise as photoconductors. Carrier mobilities are low, and the excitation energy required is often of the order of 0.5—1.5 eV (Table 7). An exception is doped poly-(vinyl chloride). Adding the leucobase of malachite green sensitized PVC which in situ developed the spectrum of the oxidized form[288]. However, direct addition of the oxidized dye form was ineffective in promoting photoconductance. Zinc sulphate in poly(vinyl chloride) and polystyrene introduces, after ultra-violet irradiation, a measure of photoconductivity[289].

Polyacetylenes[290], poly(vinylenes)[291] and conjugated polymers with heteroatoms[290, 292] often show significant levels of photoconductivity. Dye sensitization[281, 293] in some polyacetylenes and metal acetylides has been investigated. It is in the poly(diacetylenes) described earlier that large photoconductivities are observed[178, 185]. Transient measurements of the polymer with $R = -CH_2OSO_2C_6H_4CH_3$ (Fig. 15) show that carriers (both holes and electrons) have mobilities of $1 \text{ cm}^2 \cdot \text{V}^{-1} \cdot \text{s}^{-1}$, which, in the case of holes is $\approx 10^7$ times that in PNVC. Substantial trap concentrations are present, as shown by reported carrier lifetimes of $\approx 1 \mu s$, but these have been reduced by using X-ray polymerized crystals. The importance of structural regularity in good photoconductive polymers (as well as in semi-conducting polymers) is well illustrated by these high mobilities, bearing in mind that energy levels also play a major role.

The relatively low level of photoconductance required for electrophotography, together with the observation that, at least for PNVC/TNF, TNF$^-$ molecular interactions provide a photoconductive pathway, naturally leads to the question as to whether small molecule systems dispersed in a nonphotoactive polymer matrix might not provide a viable measure of photoconductivity? Other factors make such an approach advantageous. Certain polymeric analogues of organic semi-conductors are difficult to synthesise, particularly from vinyl monomers of acceptor molecules, since such monomers have by their nature low reactivity in free radical polymerization. The ability to readily vary the concentration of active molecules in a matrix can also provide valuable information on transport mechanism in terms of an adjustable mean free path. Mention has already been made of the work of Gill[276] on TNF dispersed in a polyester. Mort et al.[294] using N-isopropylcarbazole dispersed in 'Lexan' (a commercial polycarbonate) have shown the transit time for conduction has an exponential dependence on the average dopant separation. Such a distribution, related to the distribution of free paths, indicates that the hole transport occurs via hopping between N-isopropylcarbazole molecules. Kryszewski and coworkers have used this technique to study mechanistic aspects of the photoconductance of anthracene, CT complexed with chloranil and 1,3,5-trinitrobenzene, using poly-(methyl methacrylate) and polystyrene as hosts[295], and naphthalene, biphenyl, pyrene, durene, stilbene, carbazole and N-ethylcarbazole with TCNQ in a variety of polymer matrices[296]. This modern development of introducing small molecule, photo-active species into a neutral polymer matrix promises to provide a fruitful method of delving into the solid state properties of polymers. The expectation that some of these materials may also be commercially exploitable makes their continuing study of added importance.

8 References

1. Von Hippel, A. R.: Dielectrics and waves, p. 253. New York: Wiley. London: Chapman Hall, 1954
2. Mark, J. E.: Acc. Chem. Res. 7, 218 (1974)
3. Volkenstein, M. W.: Configurational statistics of polymeric chains, New York: Interscience 1963; Birshtein, T. M., Ptitsyn, O. B.: Conformation of macromolecules, New York: Interscience, 1966; Flory, P. J.: Statistical mechanics of chain molecules, New York: Interscience, 1969; Flory, P. J.: Pure Appl. Chem. 26, 309 (1971)
4. Marchal, J. Benoit, H.: J. Chim. Phys. Phys.-Chim. Biol. 52, 818 (1955); J. Polym. Sci. 23, 223 (1957)
5. Guggenheim, E. A.: Trans. Faraday Soc. 45, 714 (1949)
6. Onsager, L.: J. Am. Chem. Soc. 58, 1486 (1936)
7. Abbott, J. A., Bolton, H. C.: Trans. Faraday Soc. 48, 422 (1952)
8. Block, H., Hayes, E. F.: Trans. Faraday Soc. 66, 2512 (1970)
9. Block, H., Walker, S. M.: Chem. Phys. Lett. 19, 363 (1973)
10. Kirkwood, J. G.: J. Chem. Phys. 7, 911 (1939)
11. Fröhlich, H.: Theory of dielectrics, London: Oxford Univ. Press 1949
12. McCrum, N. G., Read, B. E., Williams, G.: Anelastic and dielectric effects in polymeric solids, London–New York: Wiley, 1967
13. Stockmayer, W. H.: Pure Appl. Chem. 15, 539 (1967)
14. Nagai, K., Ishikawa, T.: Polym. J. 2, 416 (1971); Dai, M.: Polym. J. 3, 252 (1972)
15. Svirbley, W. J., Lander, J. J.: J. Am. Chem. Soc. 67, 2189 (1945); Uchida, T., Kurita, Y., Noizum, N., Kubo, M.: J. Polym. Sci. 21, 313 (1956); Rossi, C., Magnasco, V.: J. Polym. Sci. 58, 977 (1962); Kotera, A., Suzuki, K., Matsumura, K., Nakamo, T., Oyama, T., Kambayashi, U.: Bull. Chem. Soc. Jpn. 35, 797 (1962); Mark, J. E., Flory, P. J.: J. Am. Chem. Soc. 87, 1415 (1965); 88, 3702 (1966)
16. Bak, K., Elefante, G., Mark, J. E.: J. Phys. Chem. 71, 4007 (1967)
17. Jamamoto, K., Teramoto, A., Fujita, H.: Polymer 7, 267 (1966)

18. Wetton, R. E., Williams, G.: Trans. Faraday Soc. *61*, 2132 (1965); Mark, J. E.: J. Am. Chem. Soc. *88*, 3708 (1966)
19. Riande, E.: J. Polym. Sci., Polym. Phys. Ed. *14*, 2231 (1976)
20. Uchida, T., Kurita, Y., Kubo, M.: J. Polym. Sci. *19*, 365 (1956)
21. Dasgupta, S., Smyth, C. P.: J. Chem. Phys. *47*, 2911 (1967); Sutton, C., Mark, J. E.: J. Chem. Phys. *54*, 5011 (1971); Liao, S. C., Mark, J. E.: J. Chem. Phys. *59*, 3825 (1973)
22. LeFevre, R. J. W., Sundaram, K. M. S.: J. Chem. Soc. *1962*, 1494; Kotera, A., Shima, M., Fujisaki, N., Kobayashi, T.: Bull. Chem. Soc. Jpn. *35*, 1117 (1962)
23. Mark, J. E.: J. Chem. Phys. *56*, 451 (1972)
24. Krigbaum, W. R., Roij, A.: J. Chem. Phys. *31*, 544 (1959)
25. Mark, J. E.: J. Chem. Phys. *94*, 6645 (1972)
26. Birshtein, T. M., Burshtein, L. L., Ptitsyn, O. B.: Sov. Phys.-Tech. Phys. (Engl. Transl.) *4*, 810 (1959)
27. Pohl, H. A., Zabusky, H. H.: J. Phys. Chem. *66*, 1390 (1962)
28. Debye, P., Bueche, F.: J. Chem. Phys. *19*, 589 (1951); Smith, F. H., Corrado, L. C., Work, R. N.: Polym. Prepr., Am. Chem. Soc., Div. Polym. Chem. *12*, 64 (1971); Kotera, A., Suzuki, K., Matumara, K., Shima, M., Joko, E.: Bull. Chem. Soc. Jpn. *39*, 750 (1966)
29. Baysal, B., Lawry, B. A., Yu, H., Stockmayer, W. H.: In: Dielectric properties of polymers, F. E. Karasz, ed., p. 329. New York–London: Plenum Press, 1972
30. Mark, J. E.: J. Chem. Phys. *56*, 458 (1972)
31. Yu, H., Bur, A. J., Fetters, L. J.: J. Chem. Phys. *44*, 2568 (1966); Bur, A. J., Roberts, D. E.: J. Chem. Phys. *51*, 406 (1969)
32. Burchard, W.: Makromol. Chem. *67*, 182 (1963); Schneider, N. S., Furasaki, S., Lenz, R. W.: J. Polym. Sci., Part A, *3*, 933 (1965); Milstein, J. B., Chamey, E.: Macromolecules *2*, 678 (1969); Dev, S. B., Lockhead, R. Y., North, A. M.: Discuss. Faraday Soc. *49*, 244 (1970); Plummer, H., Jennings, B. R.: Eur. Polym. J. *6*, 171 (1970); Jennings, B. R., Brown, B. L.: Eur. Polym. J. *7*, 805 (1971); Berger, M. N.: J. Macromol. Sci., Rev. Macromol. Chem. C*9*, 269 (1973); Berger, M. N., Tidswell, B. N.: J. Polym. Sci. Polym. Symp. *42*, 1063 (1973); Tsvetkov, V. N., Riumtsev, E. I., Pogodina, N. V., Shteunikova, I. N.: Eur. Polym. J. *11*, 37 (1975); Pierre, J., Marchal, E.: J. Polym. Sci., Polym. Lett. Ed. *13*, 11 (1975); Nemoto, N., Schrag, J. L., Ferry, J. D.: Polym. J. *7*, 195 (1975)
33. Tsvetkov, V. N., Riumtsev, E. I., Aliyev, F. M., Shteunikova, I. N.: Eur. Polym. J. *10*, 55 (1974)
34. Tsvetkov, V. N., Riumtsev, E. I., Aliyev, F. M., Shteunikova, I. N., Savvon, S. M.: Vysokomol. Soyedin., Ser. A: *16*, 1401 (1974), [Engl. transl.: Polym. Sci. USSR *16*, 1627 (1974)]
35. Lockhead, R. Y., North, A. M.: J. Chem. Soc., Faraday Trans. 2, *7*, 1089 (1972)
36. Ryumtsev, Ye. I., Aliyev, F. M., Tsvetkov, V. N.: Vysokomol. Soyedin., Ser. A: *17*, 1712 (1975) [(Engl. transl.): Polym. Sci. USSR *17*, 1967 (1975)]; Ryumtsev, Ye. I., Pogodina, N. V., Getmanchuk, Yu. P.: Polym. Sci. USSR *17*, 1719 (1975) [Engl. transl.: Polym. Sci. USSR *17*, 1975 (1975)]
37. Tsvetkov, V. N., Shteunikova, I. N., Ryumtsev, Ye. I., Andreyeva, L. N., Getmanchuk, Yu. P., Spirin, Yu. A., Dryagileva, R. I.: Vysokomol. Soyedin., Ser. A: *10*, 2132 (1968) [Engl. transl.: Polym. Sci. USSR *10*, 2482 (1968)]; Tsvetkov, V. N., Shteunikova, I. N., Ryumtsev, Ye. I., Getmanchuk, Yu. P.: Eur. Polym. J. *7*, 767 (1971); Lecomte, L., Marchal, E.: Eur. Polym. J. *12*, 741 (1976)
38. Abe, A.: J. Polym. Sci., Polym. Symp. *54*, 135 (1976)
39. Hayman, H. J. G., Eliezer, I.: J. Chem. Phys. *28*, 890 (1958); *35*, 644 (1961)
40. Leonard, W. J., Jernigan, R. L., Flory, P. J.: J. Chem. Phys. *43*, 2256 (1965)
41. Bates, T. W., Stockmayer, W. H.: Macromolelcules *1*, 12 (1968)
42. Bridgman, W. B.: J. Am. Chem. Soc. *60*, 530 (1938)
43. Jones, A. A., Brehm, G. A., Stockmayer, W. H.: J. Polym. Sci., Polym. Symp. *46*, 149 (1974); Jones, A. A., Stockmayer, W. H., Molinori, R. J.: J. Polym. Sci. Polym. Symp. *54*, 227 (1976)
44. Bates, T. W., Ivin, K. J., Williams, G.: Trans. Faraday Soc. *63*, 1976 (1967)

45. Wada, A.: J. Chem. Phys. *29*, 674 (1958); *30*, 328 (1958); *31*, 495 (1959); Bull. Chem. Soc. Jpn. *33*, 822 (1960); Marchal, E., Marchal, J.: C. R. Hebd. Acad. Sci. *248*, 100 (1958); *250*, 2197 (1960); Arch. Sci. *12*, 24 (1959); J. Chim. Phys. *64*, 1607 (1967); Marchal, J., Marchal, E.: Vysokomol. Soyedin., *6*, 561 (1964) [Engl. transl.: Polym. Sci. USSR *6*, 623 (1964)]

46. Wada, A., in: Polyaminoacids, polypeptides and proteins. M. A. Stahlmann, ed., p. 131; Madison, W.: Univ. of Wisconsin Press, 1961; Marchal, E., Hornick, C., Benoit, H.: J. Chim. Phys. *64*, 514 (1967); Block, H., Hayes, E. F., North, A. M.: Trans. Faraday Soc. *66*, 1095 (1969)

47. Mark, J. E.: J. Chem. Phys. *57*, 2541 (1972)

48. Mark, J. E.: J. Am. Chem. Soc. *94*, 6645 (1972)

49. Mark, J. E.: J. Polym. Sci., Polym. Phys. Ed. *11*, 1375 (1973)

50. Mark, J. E.: Polymer *14*, 553 (1973)

51. Work, R. N., Trehu, Y. M.: J. Appl. Phys. *27*, 1003 (1956)

52. Smith, F. H., Corrado, L. C., Work, R. N.: In: Dielectric properties of polymers. F. E. Karasz, ed., p. 1, New York–London: Plenum Press 1972; Corrado, L. C., Work, R. N.: J. Chem. Phys. *63*, 899 (1975)

53. Work, R. N.: J. Chem. Phys. *61*, 1006 (1974)

54. Block, H., North, A. M.: Adv. Mol. Relaxation Processes *1*, 309 (1970)

55. Brouckere, L. de, Mandel, M.: Adv. Chem. Phys. *1*, 77 (1958); North, A. M.: Chem. Soc. Rev. *1*, 49 (1972)

56. North, A. M.: Pure Appl. Chem. *39*, 265 (1974)

57. Kubo, R.: J. Phys. Soc. Jpn. *12*, 570 (1957); Lect. Theor. Phys. *1*, 120 (1958); Glarum, S. H.: J. Chem. Phys. *33*, 1371 (1960); Zwanzig, R.: Ann. Rev. Phys. Chem. *16*, 67 (1965); Cole, R.: J. Phys. Chem. *42*, 637 (1965); Cook, M., Watts, D. C., Williams, G.: Trans. Faraday Soc. *66*, 2503 (1970); Williams, G.: Chem. Rev. *72*, 55 (1972)

58. Zimm, B. H.: J. Chem. Phys. *24*, 269 (1956)

59. Rouse, P. E.: J. Chem. Phys. *21*, 1272 (1953)

60. Bueche, F.: J. Polym. Sci. *54*, 597 (1961)

61. Kuhn, W.: Helv. Chim. Acta *33*, 2057 (1950)

62. Funt, B. L., Mason, S. G.: Can. J. Res., Sect. B: *28*, 182 (1950); De Brouckere, L., Van Beek, L. K. H.: Recl. Trav. Chim. Pays-Bas *75*, 355 (1956); Veselovskii, P. F., Gandel'man, I. A.: Tr. Leningr. Politekh. Inst. *255*, 148 (1965)

63. Brouckere, L. de, Buess, D., Bock, J. de, Versheys, J.: Bull. Soc. Chim. Belg. *64*, 669 (1955)

64. Brouckere, L. de, Lecocq-Robert, A.: Bull. Soc. Chim. Belg. *70*, 549 (1961)

65. Kryszewski, M., Marchal, J.: J. Polym. Sci. *29*, 103 (1958)

66. Stockmayer, W. H., Yu, H., Davies, J. E.: Am. Chem. Soc., Div. Polym. Chem., Prepr. *4*, 132 (1963)

67. Stockmayer, W. H., Matsuo, K.: Macromolecules *5*, 766 (1972)

68. Davies, M., Williams, G., Loveluck, G. D.: Z. Elektrochem. *64*, 575 (1960)

69. Baur, M. E., Stockmayer, W. H.: J. Chem. Phys. *43*, 4319 (1965)

70. North, A. M., Phillips, P. J.: Chem. Commun. *1968*, 1340

71. Bates, T. W., Ivin, K. J., Williams, G.: Trans. Faraday Soc. *63*, 1964 (1967)

72. Williams, G.: Chem. Soc. Rev. *7*, 89 (1978)

73. Shore, J. E., Zwanzig, R.: J. Chem. Phys. *63*, 5445 (1975)

74. Cole, R. H., Cole, K. S.: J. Chem. Phys. *9*, 341 (1941)

75. Davidson, D. W., Cole, R. H.: J. Chem. Phys. *18*, 1417 (1950)

76. Fuoss, R. M., Kirkwood, J. G.: J. Am. Chem. Soc. *63*, 385 (1941)

77. Havriliak, S., Negami, S.: Polymer *8*, 161 (1967)

78. Williams, G., Watts, D. C.: Trans. Faraday Soc. *66*, 80 (1970); Williams, G., Watts, D. C., Dev, S. B., North, A. M.: Trans Faraday Soc. *67*, 1323 (1971)

79. Bergmann, K., Roberti, D. M., Smyth, C. P.: J. Phys. Chem. *64*, 665 (1960); North, A. M., Phillips, P. J.: Trans. Faraday Soc. *64*, 3235 (1968)

80. Phillips, P. J.: Ph. D. Thesis, Univ. of Liverpool, 1958; North, A. M., Phillips, P. J.: Trans. Faraday Soc. *63*, 1537 (1967)

81. Gross, G. W., Ott, H. C., Arnold, D. M.: Trans. Am. Electrochem. Soc. *74*, 193 (1938)
82. Pohl, H. A., Backsai, R., Purcell, W. P.: J. Phys. Chem. *64*, 1701 (1960); Veselovskii, P. F., Matveyev, V. K.: Vysokomol. Soyedin. *6*, 121 (1964) [Engl. transl.: Polym. Sci. USSR *6*, 1345 (1964)]
83. Brouckere, L. de, Neckel, R. Van.: Bull. Soc. Chim. Belg. *61*, 261, 452 (1952); Mikhailov, G. P., Lobanov, A. M., Platanov, M. P.: Vysokomol. Soyedin. *8*, 697 (1966) [Engl. transl.: Polym. Sci. USSR *8*, 760 (1966)]
84. Maxwell, J. C.: Lehrbuch der Elektrizität und des Magnetismus, Vol. 1, p. 328, Berlin: Springer 1883; Wagner, K. W.: Arch. Elektrotech. (Berlin) *2*, 371 (1914); Sillars, R. W.: J. Inst. Electr. Eng. *80*, 378 (1937)
85. Van-der-Tow, F., Mandel, M.: Biophys. Chem. *2*, 231 (1974)
86. Minakata, A.: Biopolymers *11*, 1569 (1972)
87. Allgen, L. G., Roswall, S.: J. Polym. Sci. *23*, 635 (1957); Johnson, G. A., Neale, S. M.: J. Polym. Sci. *54*, 229 (1961)
88. Minakata, A., Imai, N.: Biopolymers *11*, 329 (1972)
89. Mandel, M., Jenard, A.: Trans. Faraday Soc. *51*, 2158, 2170 (1963)
90. Kirkwood, J. C., Shumaker, J. B.: Proc. Nat. Acad. Sci. U.S.A. *38*, 855 (1952); Schwarz, G.: Z. Phys. Chem. (Frankfurt am Main) *19*, 286 (1959); O'Konski, C. T.: J. Phys. Chem. *64*, 605 (1960); Mandel, M.: Mol. Phys. *4*, 489 (1961); McTague, J. P., Gibbs, J. H.: J. Chem. Phys. *44*, 4295 (1966); Warashina, A., Minakata, A.: J. Chem. Phys. *58*, 4743 (1973); Pollak, M., Pohl, H. A.: J. Chem. Phys. *63*, 2980 (1975)
91. Oncley, J. L.: In: Proteins, amino acids and peptides, E. J. Cohn and J. T. Edsall, eds., p. 543, New York: Reinhold 1943; Schwan, H. P.: Adv. Biol. Med. Phys. *5*, 147 (1957); Takashima, S.: Protides Biol. Fluids, Proc. Colloq. *13*, 393 (1966)
92. Allgen, L. G.: Nature *163*, 849 (1949); O'Konski, C. T., Haltner, A. J.: J. Am. Chem. Soc. *79*, 5634 (1957); O'Konski, C. T., Yoshioka, K., Orttung, W. H.: J. Phys. Chem. *63*, 1558 (1959); Tinoco, I., Yamaoka, K.: J. Phys. Chem. *63*, 423 (1959); Ingram, P., Jerrard, H. G.: Nature *196*, 57 (1962); Takashima, S.: J. Mol. Biol. *7*, 455 (1963); Moser, P., Squire, P. G., O'Konski, C. T.: J. Phys. Chem. *70*, 744 (1966); Goluband, E. I., Nazorienko, V. G.: Biophys. J. *7*, 13 (1967)
93. Block, H., Hayes, E. F.: Chem. Commun. *1969*, 76
94. Parry-Jones, G., Gregson, M., Davies, M.: Chem. Phys. Lett. *4*, 33 (1969); Gregson, M., Parry-Jones, G., Davies, M.: Chem. Phys. Lett. *6*, 215 (1970); Brown, B. L., Parry-Jones, G., Davies, M.: J. Phys. D: *7*, 1192 (1974); Davies, M.: Acta. Phys. Pol. A *50*, 241 (1976)
95. Ullman, R.: J. Chem. Phys. *56*, 1869 (1972)
96. Brown, B. L., Parry-Jones, G.: J. Polym. Sci., Polym. Phys. Ed. *13*, 599 (1975)
97. Parry-Jones, G., Gregson, M., Krupkowski, T.: Chem. Phys. Lett. *13*, 266 (1972)
98. Block, H., Gregson, E. M., Ions, W. D., Powell, G., Singh, R. P., Walker, S. M.: J. Phys. E: *11*, 251 (1978)
99. Funt, B. L., Mason, S. G.: Can. J. Chem. *29*, 848 (1951); Hartmann, H., Jaenicke, R.: Z. Phys. Chem. (Frankfurt am Main) *6*, 220 (1956)
100. Takashima, S.: J. Phys. Chem. *74*, 4446 (1970)
101. Wendisch, P.: Kolloid Z. Z. Polym. *199*, 27 (1964)
102. Barisas, B. G.: Macromolecules *7*, 930 (1974)
103. Saito, N., Kato, T.: J. Phys. Soc. Jpn. *12*, 1393 (1957)
104. Peterlin, A., Reinhold, C.: Kolloid Z. Z. Polym. *204*, 23 (1965)
105. Block, H., Ions, W. D., Powell, G., Singh, R. P., Walker, S. M.: Proc. R. Soc. London, Ser. A: *352*, 153 (1976)
106. Block, H., Gregson, E. M., Walker, S. M.: to be published
107. Block, H., Ions, W. D., Walker, S. M.: J. Polym. Sci., Polym. Phys. Ed. *16*, 989 (1978)
108. Block, H., Goodwin, K. M. W., Gregson, E. M., Walker, S. M.: Nature *275*, 632 (1978)
109. McCall, D. W.: Relaxation in solid polymers, Nat. Bur. Stand. (U.S.), Spec. Publ. No 301, 475 (1969)
110. Williams, M. L., Landel, R. F., Ferry, J. D.: J. Am. Chem. Soc. *77*, 3701 (1955)

111. Williams, G., Watts, D. C.: In: Dielectric properties of polymers, F. E. Karasz, ed., p. 17, New York–London: Plenum Press 1972
112. North, A. M.: In: Molecular behaviour and the development of polymeric materials, A. Ledwith and A. M. North, eds., p. 368, London: Chapman and Hall 1974; North, A. M.: In: International review of science: Physical chemistry, Ser. 2, *8*, Macromolecular science, C. E. H. Bawn, ed., p. 1, London and Boston: Butterworths 1975
113. Block, H., Groves, R., Walker, S. M.: Polymer *13*, 527 (1972); Block, H., Bailey, J., Cowden, D. R., Walker, S. M.: Polymer *14*, 45 (1973); Block, H., Collinson, M. E., Walker, S. M.: Polymer *14*, 68 (1973); Block, H., Lord, P. W., Walker, S. M.: Polymer *16*, 739 (1975); Block, H., Evans, D., Walker, S. M.: Polymer *18*, 768 (1977); Block, H., Cowden, D. R., Lord, P. W., Walker, S. M.: Polymer *18*, 175 (1977)
114. Koppelmann, J., Gielessen, J.: Z. Elektrochem. *65*, 689 (1961); O'Reilly, J. M.: J. Polym. Sci. *59*, 429 (1962); Heydemann, P.: Kolloid Z. *195*, 122 (1964); Williams, G.: Trans. Faraday Soc. *60*, 1548, 1556 (1964); *61*, 1564 (1965); *62*, 1321, 2091 (1966); Saito, S., Sasabe, H., Nakajama, I., Yada, K.: J. Polym. Sci., Part A-2, *6*, 1297 (1968); Sasabe, H., Saito, S.: J. Polym. Sci., Part. A-2, *6*, 1401 (1968)
115. Block, H., Groves, R., Lord, P. W., Walker, S. M.: J. Chem. Soc., Faraday Trans. 2, *68*, 1890 (1972)
116. Hyde, P. J.: Proc. Inst. Electr. Eng. *117*, 1891 (1970)
117. Hamon, B. V.: Proc. Inst. Electr. Eng., Part 4, monograph 27 (1952)
118. Bagivov, M. A., Abasov, S. A., Malin, V. P., Jalilov, A. Ya.: J. Appl. Polym. Sci. *20*, 1069 (1976)
119. Huff, K., Müller, F. H.: Kolloid, Z. *153*, 5 (1957); Reddish, W.: Pure Appl. Chem. *5*, 723 (1962); Illers, K. H., Breuer, H.: J. Colloid Sci. *18*, 1 (1963)
120. Yemni, T., Boyd, R. H.: J. Polym. Sci., Polym. Phys. Ed. *14*, 499 (1976)
121. Baird, M. E.: Electrical properties of polymeric materials, London: Plastics Inst. Publ. 1973
122. Davies, M., Edwards, A: Trans. Faraday Soc. *63*, 2163 (1967); Davies, M., Swain, J.: Trans. Faraday Soc. *67*, 1637 (1971); Lawrie, N. G., North, A. M.: Eur. Polym. J. *9*, 345 (1973)
123. McLoughlin, J. R., Tobolsky, A. V.: J. Colloid Sci. *7*, 555 (1952)
124. Fuoss, R. M.: J. Am. Chem. Soc. *63*, 369, 378 (1941); Broens, O., Müller, F. H.: Kolloid Z. *140*, 121 (1955); *141*, 20 (1955)
125. Würstlin, F.: Kolloid Z. *134*, 135 (1953)
126. Hains, P. J., Williams, G.: Polymer *11*, 725 (1975)
127. Wada, Y., Yamamoto, K.: J. Phys. Soc. Jpn. *11*, 887 (1956); Scheiber, D. J., Mead, D. J.: J. Chem. Phys. *27*, 326 (1957); Gall, W. G., McCrum, N. G.: J. Polym. Sci. *50*, 489 (1961)
128. McCrum, N. G.: J. Polym. Sci. *54*, 561 (1961)
129. Boyd, R. H.: J. Chem. Phys. *30*, 1276 (1959); Quistwater, J. M. R., Dunell, B. A.: J. Appl. Polym. Sci. *1*, 267 (1959); Illers, K. H.: Makromol. Chem. *38*, 168 (1960)
130. Baird, M. E., Goldsworthy, G. T., Creasy, C. J.: J. Polym. Sci, Part B, *6*, 737 (1968); Polymer *12*, 159 (1971)
131. North, A. M., Reid, J. C.: Eur. Polym. J. *8*, 1129 (1972); Dev, S. B., North, A. M., Reid, J. C.: In: Dielectric properties of polymers, F. E. Karasz, ed., p. 217. New York–London: Plenum Press 1972
132. Lukomskaya, A. I., Dogadkin, B. A.: Kolloid Z. *22*, 576 (1960)
133. Van Turnhout, J.: Thermally stimulated discharge of polymer electrets, Amsterdam–New York–London: Elsevier 1975
134. Rush, K. C.: J. Macromol. Sci., Phys. B *2*, 179 (1968)
135. Bucci, C., Fieshi, R., Guidi, G.: Phys. Rev. *148*, 816 (1966)
136. Vanderschueren, J.: J. Polym. Sci., Polym. Phys. Ed. *15*, 873 (1977)
137. Fukada, E.: Adv. Biophys. *6*, 121 (1974)
138. Date, M., Takashita, S., Fukada, E.: J. Polym. Sci., Part A-2, *8*, 61 (1970); Konaga, T., Fukada, E.: J. Polym. Sci., Part A-2, *9*, 2023 (1971); Fukada, E.: Prog. Polym. Sci. Jpn. *2*, 329 (1971); Fukada, E., Furukawa, T., Baer, E., Hiltner, A., Anderson, J. M.: J. Macromol. Sci., Phys. B *8*, 475 (1973); Furukawa, T., Fukada, E.: Rep. Prog. Polym. Phys. Jpn. *16*, 457 (1973); *18*, 539 (1975); Koga, K., Kajiyama, T., Takayanagi, M.: J. Polym. Sci.,

Polym. Phys. Ed. *14*, 401 (1976); Furukawa, T., Ogiwara, K., Fukada, E.: Rep. Prog. Polym. Phys. Jpn. *19*, 533 (1976); Furukawa, T., Fukada, E.: J. Polym. Sci., Polym. Phys. Ed. *14*, 1979 (1976)

139. Fukada, E.: Proc. Int. Congr. Rheology, 15th. Vol. 3, p. 285, S. Onogi, ed., Tokyo: Univ. of Tokyo Press 1970

140. Kawai, H.: J. Appl. Phys. *8*, 975 (1969)

141. Bergman, J. G., McFee, J. H., Crane, G. R.: Appl. Phys. Lett. *18*, 203 (1971)

142. Cohen, J., Edelman, S.: J. Appl. Phys. *42*, 3072 (1971); Fukada, E., Nishiyama, K.: Jpn. J. Appl. Phys. *11*, 36 (1972); Cohen, J., Edelman, S., Vezzetti, C. F.: In: Electrets, charge storage and transport in dielectrics, Perlman, M. M., ed., p. 505, Princeton, N.J.: Electrochem. Soc. 1973; Phelan, R. J., Peterson, R. L., Hamilton, C. A., Day, G. W.: Ferroelectrics *7*, 375 (1974)

143. Murayama, N., Oikawa, T., Katto, T., Nakamura, K.: J. Polym. Sci., Polym. Phys. Ed. *13*, 1033 (1975)

144. Davis, G. T.: Piezoelectric and pyroelectric symposium, National Bureau of Standards Interagency Reports, 75–760, Broadhurst, M. G., 1975, p. 120; Baise, A. I., Lee, H., Oh, S., Salmon, R. E., Labes, H. M.: Appl. Phys. Lett. *26*, 428 (1975)

145. Hayakawa, R., Wada, Y.: Adv. Polym. Sci. *11*, 1 (1973); Wada, Y., Hayakawa, R.: Jpn. J. Appl. Phys. *15*, 2041 (1976)

146. Broadhurst, M. G., Harris, W. P., Mopsik, F. I., Malmberg, C. G.: Polym. Prepr. Am. Chem. Soc., Div. Polym. Chem. *14*(2), 820 (1973)

147. Fukada, E.: In: Electrets, charge, storage and transport in dielectrics, Perlman, M. M., ed., p. 486, Princeton, N. J.: The Electrochem. Soc. 1973

148. Putley, E. H.: Semicond. Semimetals *5*, 259 (1970)

149. Oshiki, M., Fukada, E.: Jpn. J. Appl. Phys. *15*, 43 (1976)

150. Bui, L., Shaw, H. J., Zitelli, L. T.: Electron. Lett. *12*, 393 (1976)

151. Sussner, H., Michas, D., Assfalgs, A., Hunklinger, S., Dransfield, K.: Phys. Lett. A *45*, 475 (1973); Ohigashi, H.: J. Appl. Phys. *47*, 949 (1976); Murayama, N., Nakamura, K., Obara, H., Segawa, M.: Ultrasonics *14*, 15 (1976)

152. Carpenter, R., Garner, G. M., Sear, J. F.: personal communication

153. Gallantree, H. R., Quilliam, R. M.: Marconi Rev., *39*, 189 (1976)

154. Ohigashi, H., Shigenari, R., Yokota, M.: Jpn. J. Appl. Phys. *14*, 1085 (1975)

155. Japan. Pat. 4737,244 (1968), as reported by Murayama, N., Nakamura, K., Obara, H., Segawa, M.: Ultrasonics *14*, 22 (1976)

156. Glass, A. M., McFee, J. H., Bergman, J. G.: J. Appl. Phys. *42*, 5219 (1971); Phelan, R. J., Mahler, R. J., Cook, A. R.: Appl. Phys. Lett. *19*, 337 (1971); Day, G. W., Hamilton, C. A., Peterson, R. L., Phelan, R. J., Mullen, L. D.: Appl. Phys. Lett. *24*, 456 (1974)

157. Stephens, A. W., Levine, A. W., Fech, J., Zrebiec, T. J., Cafiero, A. V., Garofalo, A. M.: Thin Solid Films *24*, 361 (1974); Garn, L. E., Sharp, E. J.: IEEE Trans. Parts, Hybrids, Packag. *10*, 208 (1974)

158. Blevin, W. R., Geist, J.: Appl. Opt. *13*, 2212 (1974)

159. Bergman, J. G., Crane, G. R., Ballman, A. A., O'Bryan, H.: Appl. Phys. Lett. *21*, 497 (1972)

160. Rembaum, A.: In: Encyclopedia of polymer science and technology, Vol. 11, ed., N. M. Bikales and J. Courad, eds., p. 318, New York: London– Sidney– Toronto: Interscience 1969

161. Rembaum, A., Somoano, R. B.: In: Electrical properties of polymers, K. C. Frisch, ed., p. 139, Westport, Conn.: Technomic 1972; Haertel, M., Kossmehl, G., Manecke, G., Wille, W., Woehrle, D., Zerpner, D.: Angew. Makromol. Chem. *29–30*, 307 (1973); Manecke, G.: Pure Appl. Chem. *1*, 155 (1974); Goodings, E. P.: Endeavour *34*, 123 (1975); Kryszewski, M.: J. Polym. Sci., Polym. Symp. *50*, 359 (1975)

162. Goodings, E. P.: Chem. Soc. Rev. *5*, 95 (1976)

163. Netherlands Pat. 6,510,986 (1966), General Electric Co. [Chem. Abs. *65*, 7382e (1966)]

164. Japan. Pat. 7427,570 (1974), S. Suzuki, H. Iwabori, F. Tamura [Chem. Abs. *81*, 128771s (1974)]

165. Norman, R. H.: Conductive rubbers and plastics, Amsterdam–London–New York: Elsevier 1970

166. Dolezalek, F. K.: In: Photoconductivity and related phenomena, J. Mort and D. M. Pai, eds, p. 26, Amsterdam–Oxford–New York: Elsevier 1976

167. Carver, G. P.: Rev. Sci. Instrum. *43*, 1275 (1972); Carver, G. P., Allgaier, R. S.: J. Non-Cryst. Solids *8–10*, 347 (1972)

168. Hankin, A. G., North, A. M.: Trans Faraday Soc. *63*, 1525 (1967)

169. Byrd, N. R., Kleist, F. K., Stamires, D. N.: J. Polym. Sci., Polym. Phys. Ed. *10*, 957 (1972)

170. Sinitskii, V. V., Myachina, G. F., Kryazhev, Yu. G., Rozenshtein, L. D.: Izv. Akad. Nauk SSSR, Ser. Khim. *21*, 972 (1972) [Engl. transl.: Bull. Acad. Sci. USSR, Div. Chem. Sci. *21*, 932 (1972)]; Sinitskii, V. V., Kazanskaya, N. V., Rozenshtein, L. D., Cherkashin, M. I., Kryazhev, Yu. G.: Izv. Akad. Nauk SSSR. Ser. Khim. *21*, 1069 (1972) [Engl. transl.: Bull. Acad. Sci.: USSR, Div. Chem. Sci. *21*, 1025 (1972)]

171. Shirakawa, H., Ikeda, S.: Polym. J. *2*, 231 (1971); Shirakawa, H., Ito, T., Ikeda, S.: Polym. J. *4*, 460 (1973); Ito, T., Shirakawa, H., Ikeda, S.: J. Polym. Sci., Polym. Chem. Ed. *12*, 11 (1974); *13*, 943 (1975)

172. Hartel, M., Kossmehl, G., Manecke, G., Wille, W., Wöhrle, D., Zerpner, D.: Angew. Makromol. Chem. *29–30*, 307 (1973)

173. Topchiev, A. V., Geiderikh, M. A., Davydov, B. E., Kargin, V. A., Krentsel, B. A., Kustanovitch, I. M., Polak, L. S.: Dokl. Akad. Nauk SSSR *128*, 312 (1959); Geiderikh, M. A., Davydov, B. E., Krentsel, B. A., Kustanovich, I. M., Polak, L. S., Topchiev, A. V., Voitenko, R. M.: Mezhdunar. Simp. Makromol. Khim., Dokl. Autoreferaty *3*, 85 (1960)

174. Rembaum, A., Moacanin, J., Pohl, H. A.: Prog. Dielectr. *6*, 41 (1965)

175. Nesmeyanov, A. N., Rybinskaia, M. I., Slonimskii, G. L.: Vysokomol. Soyedin. *2*, 526 (1960), [Engl. abstract: Polym. Sci. USSR *2*, 477 (1961)]

176. Mason, J. W., Pohl, H. A., Hartman, R. D.: J. Polym. Sci., Part C *17*, 187 (1967)

177. Hartmann, R. D., Pohl, H. A.: J. Polym. Sci., Part A-1, *6*, 1135 (1968); Pohl, H. A., J. Biol. Phys. *2*, 113 (1974)

178. Baughman, R. H.: In: Contemporary topics in polymer science, Vol. 2, p. 205, E. M. Pearce and J. R. Shaefgen, eds., New York: Plenum 1977

179. Shermann, W., Wegner, G.: Makromol. Chem. *154*, 35 (1972); *175*, 667 (1974)

180. Wegner, G.: Chimia *28*, 475 (1974)

181. Hädicke, E., Mez, E. C., Krauch, C. H., Wegner, G., Kaiser, J.: Angew. Chem., Int. Ed. Engl. *10*, 266 (1971); Shultz, J. M.: J. Mater. Sci. *11*, 2258 (1976)

182. Sauteret, C., Hermann, J. P., Frey, R, Pradere, F., Ducuing, J., Baughman, R. H., Chance, R. R.: Phys. Rev. Lett. *36*, 956 (1976)

183. Baughman, R. H., Gleiter, H., Sendfield, N.: J. Polym. Sci., Polym. Phys. Ed. *13*, 1871 (1975)

184. Bloor, D., Ando, D. J., Preston, F. H., Stevens, G. C.: Chem. Phys. Lett. *24*, 407 (1974)

185. Chance, R. R., Baughman, R. H., Reucroft, P. J., Takashi, K.: Chem. Phys. *13*, 181 (1976)

186. Bloor, D., Preston, F. H., Ando, D. J.: Chem. Phys. Lett. *38*, 33 (1976); Lochner, K., Reimer, B., Bässler, H.: Chem. Phys. Lett. *41*, 388 (1976); Eckhardt, D. J., Müller, H., Tylicki, J., Chance, R. R.: J. Chem. Phys. *65*, 4311 (1976)

187. Exarhas, G. J., Risen Jr., W. M., Baughman, R. H.: J. Am. Chem. Soc. *98*, 481 (1976); Baughman, R. H., Chance, R. R.: J. Polym. Sci., Polym. Phys. Ed. *14*, 2037 (1976)

188. Baughman, R. H., Chance, R. R., Cohen, M. J.: J. Chem. Phys. *64*, 1886 (1976); Baughman, R. H., Chance, R. R.: J. Polym. Sci., Polym. Phys. Ed. *14*, 2019 (1976)

189. Smith, R. D., Wyatt, J. R., De Corpo, J. J., Saalfield, F. E., Moran, M. J., MacDiarmid, A. G.: Chem. Phys. Lett. *41*, 362 (1976)

190. Cohen, J., Garito, A. F., Heeger, A. J., MacDiarmid, A. G., Mikulski, C. M., Saran, M. S., Kleppinger, J.: J. Am. Chem. Soc. *98*, 3844 (1976)

191. Walatka, V. V., Labes, M. M., Perlstein, J. H.: Phys. Rev. Lett. *31*, 1139 (1973); Hsu, C. H., Labes, M. M.: J. Chem. Phys. *61*, 4640 (1974); Bright, A. A, Cohen, M. J., Garito, A. F., Heeger, A. J., Mikulski, C. M., MacDiarmid, A. G.: Appl. Phys. Lett. *26*, 612 (1975)

192. Geserich, H. P., Pintschovius, L.: Festkörperprobleme *16*, 65 (1976)

193. Greene, R. L., Street, G. B.: Proc. of the NATO-ASI on the chemistry and physics of one-dimensional metals, Bolzano, Italy, 1976, H. J. Keller, ed., New York–London: Plenum Press 1977

194. Greene, R. L., Street, G. B., Suter, L. J.: Phys. Rev. Lett. *34*, 577 (1975)
195. Civiak, R., Junker, W., Elbaum, C., Kao, H. I., Labes, M. M.: Solid State Commun. *17*, 1573 (1975); Civiak, R., Elbaum, C., Junker, W., Gough, C., Kao, H. I., Nichols, L. F., Labes, M. M.: Solid State Commun. *18*, 1205 (1976); Azevedo, L. J., Clark, W. G., Deutscher, G., Greene, R. L., Street, G. B., Suter, L. J.: Solid State Commun. *19*, 197 (1976)
196. Kahlert, H., Seeger, K.: Electrical properties of polysulphur nitride $(SN)_x$, Proc. 13th Int. Conf. Phys. Semicond., Rome, Aug. 1976
197. Perlstein, J. H.: Angew. Chem., Internat. Ed. Engl. *16*, 519 (1977)
198. Eley, D. D.: J. Polym. Sci., Part C, *17*, 78 (1967)
199. Schneider, A. A., Greatbach, W., Mead, R.: 9th Int. Power Sources Symp., Brighton, U. K., 1974, Paper No. 30
200. Kuoesel, R., Gebus, B., Roth, J. P., Parrod, J.: Bull. Soc. Chim. France *1969*, 294
201. Hermann, A. M., Rembaum, A.: J. Polym. Sci., Part C, *17*, 107 (1967)
202. Litt, M. H., Summers, J. W.: J. Polym. Sci., Polym. Chem. Ed. *11*, 1339 (1973)
203. Naarmann, H.: Angew. Chem. *81*, 871 (1969); German Pat. 1953898 (1971) [Chem. Abs. *75*, 37498t (1971)]; German Pat. 1953899 (1971) [Chem. Abs. *75*, 64634r (1971)]
204. Sulzberg, T., Cotter, R. J.: Macromolecules *1*, 554 (1968); J. Polym. Sci., Part A-1, *8*, 2747 (1970)
205. Hoegl, H.: J. Phys. Chem. *69*, 755 (1965)
206. Suchanski, M. R.: J. Electrochem. Soc. *123*, 181C (1976)
207. Etemad, S., Penney, T., Engler, E. M., Scott, B. A., Seiden, P. E.: Phys. Rev. Lett. *14*, 741 (1975)
208. Lupinski, J. H., Kopple, K. D.: Science *146*, 1038 (1964); Lupinski, J. H., Kopple, K. D., Hertz, J. J.: J. Polym. Sci., Part C, *16*, 1561 (1967)
209. Naomori, H., Hatano, M., Kambara, S.: Kogyo Kagaku Zasshi *67*, 1608 (1964)
210. Boniface, O. W., Braithwaite, M. J., Eley, D. D., Evans, R. G., Pithig, R., Willis, M. R.: Disc. Farad. Soc. *51*, 131 (1971)
211. Goodings, E. P.: Discuss. Faraday Soc. *51*, 157 (1971)
212. Hadek, V., Noguchi, H., Rembaum, A.: Macromolecules *4*, 494 (1971)
213. Rembaum, A., Hermann, A. M., Haak, R.: J. Polym. Sci., Part A-1, *6*, 1955 (1968); Rembaum, A., Baumgarten, W., Eisenberg, A.: J. Polym. Sci., Part B, *6*, 159 (1968)
214. Rembaum, A., Hermann, A. M., Stewart, F. E., Gutman, F.: J. Phys. Chem. *73*, 513 (1969)
215. Yen, S. P. S., Somoano, R., Hadek, V., Cuellar, E., Washington, E., Rembaum, A.: Polym. Prepr., Am. Chem. Soc., Div. Polym. Chem. *17*, 321 (1976)
216. Factor, A., Heinsohn, G. E.: J. Polym. Sci., Part B, *9*, 289 (1971)
217. Bruce, J. M., Herson, J. R.: Polymer *8*, 619 (1967)
218. Somoano, R., Yen, S. P. S., Rembaum, A.: J. Polym. Sci., Part B, *8*, 467 (1970)
219. Hermann, A. M., Yen, S. P. S., Rembaum, A., Landell, R. F.: J. Polym. Sci., Part B, *9*, 627 (1971)
220. Nakatani, K., Saka, T., Tsubomura, H.: Bull. Chem. Soc. Jpn. *48*, 2205 (1975)
221. Manecke, G., Kautz, J.: Makromol. Chem. *172*, 1 (1973)
222. Block, H., Cowd, M. A., Walker, S. M.: Polymer *18*, 781 (1977)
223. Uena, Y., Masuyama, Y., Okawara, M.: Chem. Lett., 603 (1975); Pittman, C. U., Narita, M., Liang, Y. F.: Macromolecules *9*, 360 (1976)
224. Hertler, W. R.: J. Org. Chem. *41*, 1412 (1976)
225. Dewar, M. J. S., Talati, A. M.: J. Am. Chem. Soc. *85*, 1874 (1963); *86*, 1592 (1964)
226. Akiyama, Y., Mizutani, H.: J. Phys. Soc. Jpn. *26*, 1128 (1969)
227. Epstein, A., Wildi, B. S.: J. Chem. Phys. *32*, 324 (1960); Wildi, B. S., Katon, J. E.: J. Polym. Sci. Part A-2, *2*, 4709 (1964)
228. Norrell, C. J., Pohl, H. A., Thomas, M., Berlin, K. D.: J. Polym. Sci., Polym. Phys. Ed. *12*, 913 (1974)
229. Hamann, C., Schmidt, H.: Plaste Kautsch. *16*, 85 (1969)
230. Meyer, G., Wöhrle, D.: Makromol. Chem. *175*, 714 (1974)
231. Levina, S. D., Labanova, K. P., Berlin, A. A., Sherle, A. I.: Dokl. Akad. Nauk SSSR *145*, 602 (1962)

232. Garito, A. F., Haeger, A. J.: Acc. Chem. Res. 7, 232 (1974)
233. Sasaki, Y., Walker, L. L., Hurst, E. L., Pitmann, C. U.: J. Polym. Sci., Polym. Chem. Ed.
     11, 1213 (1973); Pittmann, C. U., Sasaki, Y., Grube, P. L.: J. Macromol. Sci., Chem., *8*,
     923 (1974)
234. Pittman, C. U., Sasaki, Y., Mukherjee, T. K.: Chem. Lett. 383 (1975)
235. Cowan, D. O., Park, J., Pittman, C. U., Sasaki, Y., Mukherjee, T. K., Diamond, N. A.:
     J. Am. Chem. Soc. *94*, 5110 (1972)
236. Pitman, C. U., Surynarayanan, B.: J. Am. Chem. Soc. *96*, 7916 (1974)
237. Krogmann, K.: Angew. Chem., Internat. Ed. Engl. *8*, 35 (1969)
238. Minot, M. J., Perlstein, J. H.: Phys. Rev. Lett. *26*, 371 (1971)
239. Yoffe, A. D.: Chem. Soc. Rev. *5*, 51 (1976)
240. Comés, R., Lambert, M., Lannois, H., Zeller, H. R.: Phys. Rev., Sect. B, *8*, 571 (1973);
     Phys. Status Solidi B: *58*, 587 (1973)
241. Holliday, A. K., Hughes, G., Walker, S. M.: The chemistry of carbon: Organometallic chem-
     istry, Chap. 13, Pergamon Texts in Inorganic Chemistry, Vol. 6, A. F. Trotman-Dickinson,
     ed., Oxford–New York–Toronto–Sydney–Braunschweig: Pergamon Press 1973
242. Falardean, E. R., Foley, G. M. T., Zeller, H. R., Vogel, F. L.: J. Chem. Soc., Chem. Commun.
     1977, 389
243. MacDiarmid, A. G.: Am. Chem. Soc., Div. Org. Coat. Plast. Chem., Pap. *38*, 625 (1978)
244. MacDiarmid, A. G.: Am. Chem. Soc., Div. Org. Coat. Plast. Chem. Pap. *38*, 631 (1978)
245. Report in: Chem. Eng. News *56*, 19 (1978)
246. German Pat. 1233265 (1967), P. M. Cassiers, J. M. Nyss, R. M. Hart and J. S. Williams;
     Japan Pat. 4-17907 (1967), K. Morimoto and A. Inami
247. German Pat. 1068115 (1957), H. Hoegl, O. Sus, W. Neugebauer [Chem. Abs. *55*, 20742a
     (1961)]; U.S. Pat. 3 162 532 (1957), H. Hoegl and W. Neugebauer
248. Rose, A.: Concepts in photoconductivity, New York: Wiley Interscience, 1963
249. Sharp, J. H., Smith, M.: In: Physical chemistry, Chap. 8, Vol. 10, New York: Academic
     Press, 1969
250. Photoconductivity and related phenomena, J. Mort, and D. M. Pai, eds., Amsterdam–
     Oxford–New York: Elsevier 1976
251. Photoconductivity in polymers; an interdisciplinary approach, A. V. Patsis and D. A. Seanor,
     eds., Westpoint, Conn.: Technomic Pub. 1976
252. Vannikov, A. V.: Sov. Phys.-Solid State (Engl. Transl.) *9*, 755 (1965)
253. Reiser, A., Lock, M. W. B., Knight, J.: Trans Faraday Soc. *65*, 2168 (1969)
254. Ranicur, J. H., Flemming, R. J.: J. Polym. Sci., Part A-2, *10*, 1321 (1972)
255. Martin, E. H., Hirsch, J.: Solid State Commun. *7*, 783 (1969); J. Non-Cryst. Solids *4*, 133
     (1970); J. Appl. Phys. *43*, 1001 (1972)
256. Davies, D. K.: J. Phys. D: *5*, 162 (1972)
257. Matsumoto, S., Yahagi, K.: J. Appl. Phys. *12*, 930 (1973)
258. Hayashi, K., Yoshino, K., Innishi, Y.: Jpn. J. Appl. Phys. *12*, 754 (1973)
259. Regensburger, P. J.: Photochem. Photobiol. *8*, 429 (1968)
260. Pai, D. M.: J. Chem. Phys. *52*, 2285 (1970)
261. Szymanski, A., Labes, M. M.: J. Chem. Phys. *50*, 3568 (1969)
262. Mort, J., Lakatos, A. I.: J. Non-Cryst. Solids *4*, 117 (1970)
263. Hermann, A. M., Rembaum, A.: J. Appl. Phys. *37*, 3642 (1966)
264. Hermann, A. M.: In: Electrical properties of polymers, K. C. Frisch and A. Patsis, eds.,
     p. 103, Westport, Conn. Technomic: 1972
265. Seki, H., Gill, W. D.: Proc. Second Int. Conf. on Conductivity in Low Mobility Materials,
     Eilat 1971, p. 409, London: Taylor and Francis, 1971
266. Gill, W. D.: J. Appl. Phys. *43*, 5033 (1972)
267. Mort, J., Emerald, R. L.: J. Appl. Phys. *45*, 175 (1974)
268. Emerald, R. L., Mort, J.: J. Appl. Phys. *45*, 3943 (1974)
269. Lakatos, A. I., Mort, J.: Phys. Rev. Lett. *21*, 1444 (1968)
270. Hermann, A. M., Rembaum, A.: J. Polym. Sci., Part B, *5*, 445 (1967)
271. Weiser, G.: J. Appl. Phys. *43*, 5028 (1972)

272. Weiser, G.: Phys. Status Solidi *18*, 347 (1973)
273. Rocklitz, J.: In: Current problems in electrophotography, W. F. Berg and K. Hauffe, eds., p. 244, Berlin: de Gruyter 1972
274. Schaffert, R. M.: IBM J. Res. Dev. *15*, 75 (1971)
275. Fox, S. J.: Electrophotography, Second Int. Conf. Society of Photographic Scientists and Engineers Publication, Washington, D. C., 1974, p. 170
276. Gill, W. D.: Proc. 5th int. conf. on amorphous and liquid semiconductors, Garmisch-Partenkirchen, 1973, p. 901, London: Taylor and Francis
277. Frenkel, J.: Phys. Rev. *54*, 647 (1938); Hill, R. M.: Philos. Mag. *23*, 59 (1971)
278. Gill, W. D.: Polymeric photoconductors. In: Photoconductivity and related phenomena, Chap. 8, J. Mort and D. Pai, eds. Amsterdam–Oxford–New York: Elsevier 1976
279. Holstein, T. D.: Ann. Phys. (N.Y.) *8*, 343 (1959); Bagley, B. G.: Solid State Commun. *8*, 345 (1970); Seki, H.: Proc. 5th int. conf. on amorphous and liquid semi-conductors, Garmisch-Partenkirchen, 1973, p. 1015. London: Taylor and Francis 1974
280. Ikeda, M., Morimoto, K., Murrakami, Y., Sato, H.: Jpn. J. Appl. Phys. *8*, 759 (1969)
281. Mylnikov, V., Terenin, A.: Mol. Phys. *8*, 387 (1964); J. Polym. Sci., Part C, *16*, 3655 (1968)
282. Block, H., Bowker, S. M., Walker, S. M.: Polymer *19*, 531 (1978)
283. Pittman, C. U., Grube, P. L.: J. Appl. Polym. Sci. *18*, 2264 (1974)
284. Belg. Pat. 812434 (1973), W. W. Limburg and D. A. Seanor
285. Okamoto, K., Kato, K., Murao, K., Kusabayashi, S., Mikawa, H.: Bull. Chem. Soc. Jpn. *46*, 2883 (1973)
286. Morimoto, K., Inami, A.: Kogyo Kagaku Zasshi *67*, 1938 (1964)
287. Williams, D. J., Limberg, W. W., Pearson, J. M., Goedde, A. O., Yanus, J. F.: J. Chem. Phys. *62*, 1501 (1975)
288. Mehl, W., Wolf, N. E.: J. Phys. Chem. Solids *25*, 1221 (1964)
289. Kryszewski, M., Skorko, M.: J. Polym. Sci., Part C, *4*, 1401 (1963)
290. Mylnikov, V. S., Sladkov, A. M., Kudryavtsev, Y. P., Luneva, L. K., Korshak, V. V., Terenin, A. N.: Dokl. Akad. Nauk SSSR *144*, 840 (1962)
291. Bock, F.: Berichte der Bunsengesellschaft für Physikalische Chemie *68*, 558 (1964)
292. Mylnikov, V. S., Putzeiko, E. K., Terenin, A. N.: Dokl. Akad. Nauk. SSSR *149*, 897 (1963)
293. Mylnikov, V. S., Terenin, A. N.: Dokl. Akad. Nauk SSSR *153*, 1381 (1963)
294. Mort, J., Pfister, G., Grammatica, S.: Solid State Commun. *18*, 693 (1976)
295. Degorski, A., Kryszewski, M.: J. Chem. Soc., Faraday Trans. 2, *71*, 1513 (1975)
296. Kryszewski, M., Wojciechowski, P.: Mol. Cryst. Liq. Cryst. *32*, 183 (1976)

Received March 16, 1979
W. Kern (editor)

Author Index Volumes 1–33

Polymer Bulletin

Editors:

Prof. H.-J. Cantow
Institute of Macromolecular
Chemistry
University of Freiburg
Stefan-Meier-Strasse 31
D-78 Freiburg/Germany

Prof. J.P. Kennedy
Dept. of Polymer Science
The University of Akron
Akron, OH 44325/USA

Prof. T. Saegusa
Dept. of Synthetic Chemistry
Kyoto University
Kyoto, 606 Japan

The articles are to be sent to one of the editors or to
Springer-Verlag Berlin Heidelberg New York

Polymer Bulletin

Preface

To cope with the rapid progress of polymer science, a new
journal is now published characterized by emphasis on rapid
publication of papers containing a most concise description of
results.

The character of the new journal is between the purely archival
journal of full papers and the so-called "letter journals" consisting exclusively of short communications.

Springer International

The journal consists of one volume a year, published in 12
issues.

Subscription information upon request.

A. Knop, W. Scheib
Chemistry and Application of Phenolic Resins

1978. 111 figures, 87 tables. XIV, pages
(Polymers/Properties and Applications,
Volume 3)
ISBN 3-540-09051-7

Contents:
Historical and Economical Development of
Phenolic Resins. – Raw Materials. – Reaction
Mechanisms. – Resin Production. – Physio-
logy and Environmental Protection. – Ana-
lytic Methods. – Degradation of Phenolic
Resins by Heat, Oxygen and High Energy
Radiation. – Modified and Thermal-
Resistant Resins. – Composite Wood Mate-
rials. – Molding Compounds. – Heat and
Sound Insulation Materials. – Industrial
Laminates and Paper Impregnation. –
Coatings. – Foundry Resins. – Abrasive
Materials. – Friction Materials. – Phenolic
Resins in Rubbers and Adhesives. – Phenolic
Antioxidants. – Other Applications. – Index.

Springer-Verlag
Berlin
Heidelberg
NewYork

Advances in Polymer Science
Fortschritte der Hochpolymeren-Forschung

Editors: H.-J. Cantow, G. Dall'Asta, K. Dušek,
J. D. Ferry, H. Fujita, M. Gordon, W. Kern,
G. Natta, S. Okamura, C. G. Overberger,
T. Saegusa, G. V. Schulz, W. P. Slichter,
J. K. Stille

Volume 30
Physical Chemistry

1979. 113 figures, 32 tables. III, 231 pages
ISBN 3-540-09199-8

Contents:
E. Stahl, V. Brüderle: *Polymer Analysis by
Thermofractography*. S. Bywater: *Preparation
and Properties of Star-Branched Polymers*.
Z. Tuzar, P. Kratchovíl, M. Bohdanecký:
*Dilute Solution Properties of Aliphatic Poly-
amides*. W. Welte, W. Kreutz: *A General Theory
for the Evaluation of X-Ray Diagrams of Bio-
membranes and Other Lamellar Systems*.

Volume 31
Chemistry

1979. 45 figures, 61 tables. III, 179 pages
ISBN 3-540-09200-5

Contents:
H. Yuki, K. Hatada: *Stereospecific Polymeriza-
tion of Alpha-Substituted Acrylic Acid Esters*.
M. Biswas, N. C. Maity: *Molecular Sieves as
Polymerization Catalysts*. L. Szegö: *Modified
Polyethylene Terephthalate Fibers*. N. A. Plate,
O. V. Noah: *A Theoretical Consideration of the
Kinetics and Statistics of Reactions of Func-
tional Groups of Macromolecules*.

Volume 32
Polymers
Syntheses/Reactivities/
Properties

1979. 56 figures, 44 tables. Approx. 180 pages
ISBN 3-540-09442-3

Contents:
S. Cesca, A. Priola, M. Bruzzone: *Synthesis
and Modification of Polymers Containing a
System of Conjugated Double Bonds*.
V. T. Stannett et al.: *Recent Advances in
Membrane Science and Technology*. G. Henrici-
Olivé, S. Olivé: *Molecular Interactions and
Macroscopic Properties of Polyacrylonitrile and
Model Substances*.